Revision for CHEMISTRY *GCSE*

• *with answers* •

B EARL

is Head of Chemistry and Information Technology at St Aidan's Church of England High School, Harrogate

L D R WILFORD

is Head of Science Faculty at St Aidan's Church of England High School, Harrogate

JOHN MURRAY

Other titles available in this series:

Revision for English	Key Stage 3 *with answers*	0 7195 7025 5
Revision for French	GCSE *with answers and cassette*	0 7195 7306 8
Revision for German	GCSE *with answers and cassette*	0 7195 7309 2
Revision for History	GCSE Modern World History	0 7195 7229 0
Revision for Maths	Levels 3–8 Key Stage 3 and Intermediate GCSE *with answers*	0 7195 7083 2
Revision for Maths	GCSE *with answers*	0 7195 7461 7
Revision for Science	Key Stage 3 *with answers*	0 7195 7249 5
Revision for Science	Key Stage 4 *with answers*	0 7195 7422 6
Revision for Spanish	GCSE *with answers and cassette*	0 7195 7394 7

■

First published in 1998 by
John Murray (Publishers) Ltd
50 Albemarle Street
London W1X 4BD

Layouts by Amanda Hawkes.
Illustrations by Barking Dog Art; Philip Ford; Linden Artists.
Typeset in 11/13pt Garamond Book by Wearset, Boldon, Tyne and Wear.
Printed and bound in Great Britain by St Edmundsbury Press Ltd, Bury St Edmunds.

A CIP catalogue entry for this title can be obtained from the British Library.

ISBN 0 7195 7632 6

Contents

Acknowledgements

The authors would like to thank Irene, Katharine, Michael and Barbara for their never-ending patience and encouragement during the production of this textbook. In addition, thanks is given to Mr Dennis Richards, Headteacher, St Aidan's Church of England High School, Harrogate, for his support. Also we wish to thank the editorial staff at John Murray, Jane Roth and Jane Fransella, for all their hard work and support.

The Examination Boards

This book was written to cover all GCSE Chemistry syllabuses. The authors and publishers are grateful to the following examination boards for kind permission to reproduce past examination questions. The answers supplied are written by the authors. The examination boards have not approved the answers and bear no responsibility for their accuracy. The addresses given below are those of the examination boards from whom specific syllabuses can be obtained.

Welsh Joint Education Committee (WJEC),
245 Western Avenue,
Cardiff,
CF5 2YX

Northern Examinations and Assessment Board (NEAB),
Devas Street,
Manchester,
M15 6EX

Southern Examination Group (SEG),
Stag Hill House,
Guildford,
Surrey,
GU2 5XJ

Midland Examining Group (MEG),
Syndicate Buildings,
1 Hills Road,
Cambridge,
CB1 2EU

London Examinations, A division of
EDEXCEL Foundation (London),
Stewart House,
32 Russell Square,
London,
WC1B 5DN

Northern Ireland Council for the Curriculum,
Examinations and Assessment (NICCEA),
Clarendon Dock,
29 Clarendon Road,
Belfast,
BT1 3BG

Photo acknowledgements

The publishers are grateful to the following for permission to reproduce copyright photographs: p.58 Mary Evans Picture Library; p.78 (both) Claude Nuridsany and Marie Perennou/Science Photo Library; p.79 (top left) David A. Ponton/Planet Earth; (lower left) J. R. Bracegirdle/Planet Earth; (top right) T. Stewart/ZEFA; (lower right) Lythgoe/Planet Earth.

Introduction

During the last few years, you have gained a great deal of important knowledge and understanding from the National Curriculum - Science (Chemistry). You have probably forgotten some of the work. This work must be revised and relearned for your examinations. Unfortunately, revision is hard work and it is easy to avoid. Some of the most common ways are shown in the cartoons below.

First, get organised!

Plan

It is always a good idea to plan your revision. The plan below is an example of a revision timetable. Each box represents 2 or 3 hours of work. There is one box for a school day, and two for weekend and holiday days. Make up something like this for your own revision, and stick it on your wall.

Study/Revision planner

1. Plan all your revision before you begin it.
2. Each box represents 2 or 3 hours of work.
3. Try to keep to your plan, but don't be afraid to change it.

Mon	Tues	Wed	Thurs	Fri	Sat am	Sat pm	Sun am	Sun pm
Section 1 1.1–1.4	Section 1 1.5–1.8	Section 1 1.9–1.11	Try the exam questions	20 March	Section 2 2.1–2.4	Section 2 2.5–2.8	Day off	Day off
Section 2 2.9–2.12	Section 2 2.13–2.15	Try the exam questions	etc	27 March				
				3 April				
				10 April				

Be an active learner

Try to be an active learner. Rather than just reading through notes, make sure that you write or draw or underline as you read through them. Make summaries and answer questions. Use spider diagrams and revision summaries (examples of these are given below).

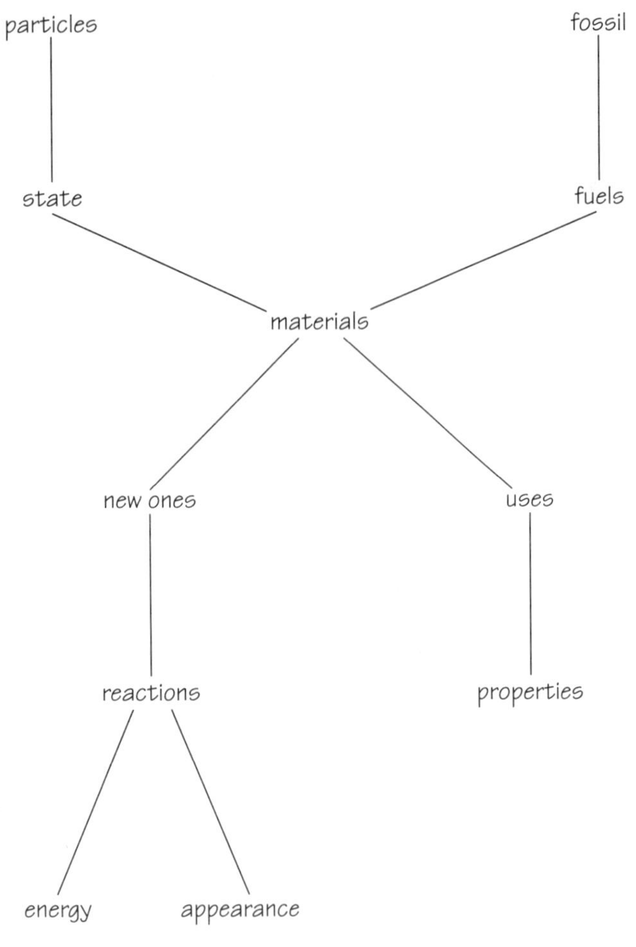

Example of a spider diagram

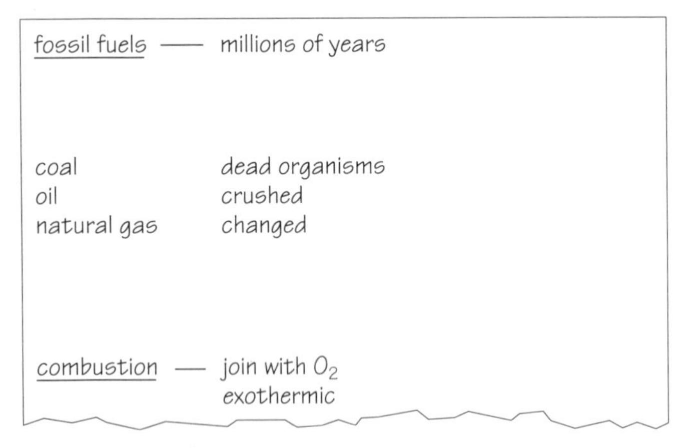

Example of a revision summary

Next, use this book!

This book has been written to help you prepare for your GCSE Chemistry examination - whether as separate Science: Chemistry, or as the Chemistry component of a Double Award (Coordinated) course.

This book will be of most use to you as you come to the end of your Chemistry course, especially when you are ready to start with your revision. Of course this book can be useful in other ways, such as when you are preparing for tests or internal examinations during your GCSE course.

Your teacher will suggest which level you should aim for, foundation or higher. Also it is very important for you to obtain a copy of the syllabus you are studying from your teacher.

1 This book has been written in three sections. Read each section carefully and make sure you understand the sense of the work. Do the quick questions as you go along. Answers to numerical questions are given on page 131.
2 To help draw attention to the more important words, the first time scientific terms are used they are printed in **bold**. These words and phrases are included in a glossary on pages 120–125. This glossary can be used as a self-test or as a simple reference section.
3 Examination questions are given at the end of each section so you can test your knowledge and understanding of the work covered. The level of each question (foundation or higher) is indicated at the end of the question. Questions are from GCSE Science: Chemistry papers, unless it is stated 'Double Award' or 'Coordinated' at the end of the question. Check your answers with those on pages 126–130 as you go along (see your teacher about anything you still do not understand). There is a reference set of data tables at the back of the book to help you quickly find other useful information.
4 Try to read and re-read the relevant topics as often as you can so that you become familiar with the ideas and learn more effectively.

Section ONE

1.1 Solids, liquids and gases

A **solid**, at a given temperature, has a definite volume and shape. Solids usually increase slightly in size when heated (**expansion**) and usually decrease in size if cooled (**contraction**).

A **liquid**, at a given temperature, has a fixed volume but it will take up the shape of any container into which it is poured. Like a solid it is also slightly affected by changes in temperature, for example, the expansion of alcohol or mercury in a thermometer is used to measure temperature changes.

A **gas**, at a given temperature, has neither a definite shape nor volume. It will take up the shape of any container into which it is placed and will spread out evenly within the containing vessel.

Unlike solids and liquids, the volumes of gases are affected quite markedly by changes in temperature.

Liquids and gases, unlike solids, are relatively **compressible**. This means the volume can be reduced by the application of pressure. Gases are much more compressible than liquids.

The kinetic theory of matter

The **kinetic theory** helps to explain the way in which **matter** behaves.

The main points of the theory are:

- All matter is made up of tiny, moving particles, invisible to the naked eye. Different substances have different types of particles (atoms, molecules or ions) which have different sizes.
- The particles move all the time. The higher the temperature, the faster they move on average.
- Heavier particles move more slowly than lighter ones at a given temperature.

The kinetic theory can be used as a scientific model to explain the properties of the three **states of matter**.

Explaining the states of matter

In a solid the particles attract one another. There are **bonds** between the particles which hold them close together. The particles are arranged in a regular manner which explains why many solids form crystals. Crystals have regular arrangements of particles within them. The particles have little freedom of movement and are only able to vibrate about a fixed position.

Figure 1.1.1 Model of a chrome alum crystal

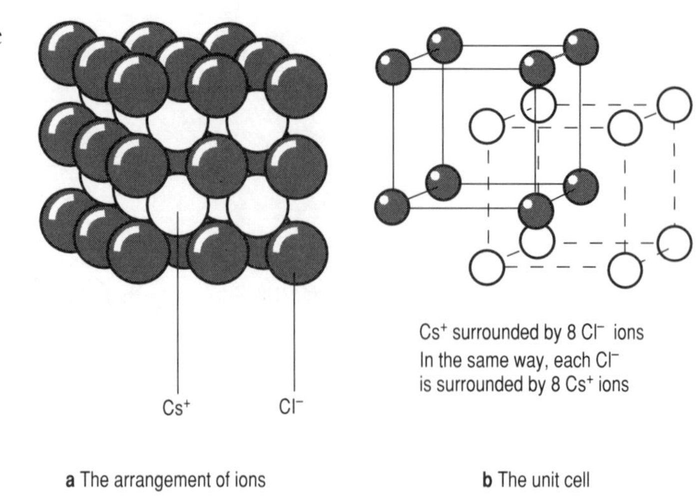

Figure 1.1.2 Structure of caesium chloride

Studies using X-ray crystallography have confirmed how the particles are arranged in crystal structures. When crystals of a pure substance form under a given set of conditions it is found that the particles present are always packed in the same way. However, the particles may be packed in different ways in crystals of different substances. For example, caesium chloride has its particles arranged in the way shown in Figure 1.1.2.

The attraction between the particles in a liquid is weaker than in a solid. The particles are still quite close together, but they are moving around in a random way and collide. The particles in the liquid form of a substance have more energy on average than the particles in the solid form of the same substance and so collide with one another quite often.

In a gas the particles are relatively far apart. They are free to move anywhere within the container in which they are held. They are moving randomly at very high velocities, much more rapidly than in a liquid. They move in straight lines unless they hit each other or the walls of their container.

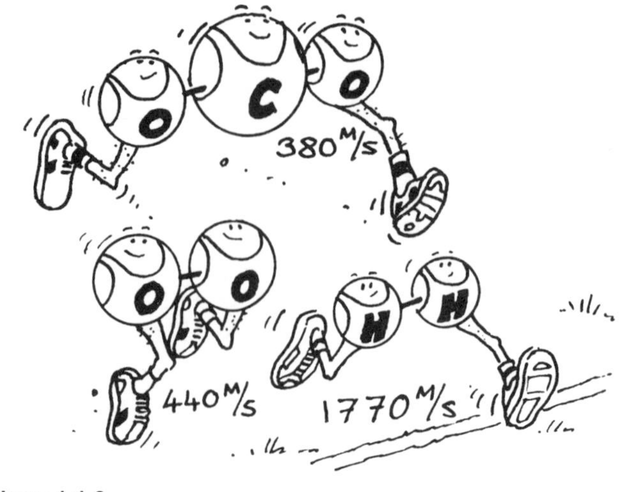

Figure 1.1.3

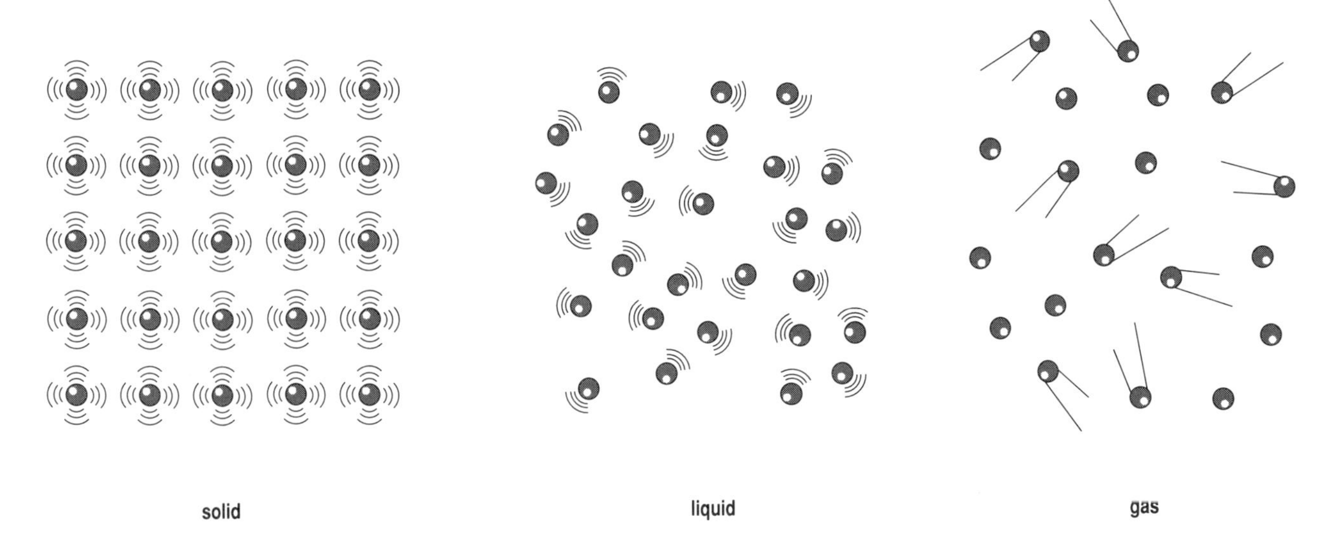

Figure 1.1.4

They collide with each other, but less often than in a liquid because they are separated more, and with the walls of the container. They exert virtually no forces of attraction on each other because they are relatively far apart. Such forces, however, are significant when the particles are close to each other. If they did not exist we could not have solids or liquids.

Diffusion – evidence for moving particles

When you walk past the cosmetics counter of a department store you can usually smell the perfumes. For this to happen, gas particles must be leaving the open perfume bottles and spreading out into the air. This spreading out of a gas is called **diffusion** and it takes place in a haphazard and random way.

Gases diffuse at different rates. This can be shown by carrying out the experiment shown in Figure 1.1.5.

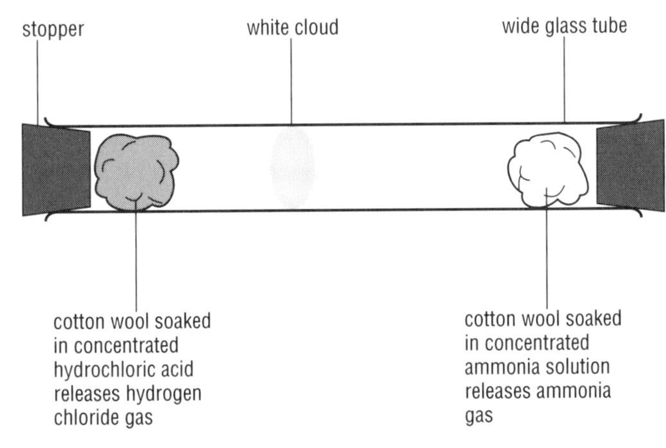

Figure 1.1.5 Diffusion of hydrogen chloride and ammonia

The white cloud of ammonium chloride shows where the two gases meet and react. It forms in this position because the ammonia particles are lighter than the hydrogen chloride particles (released from the hydrochloric acid) and so move faster. Lighter particles move faster than heavier ones at a given temperature.

Diffusion also takes place in liquids but it is a much slower process. This is because the particles of a liquid move much more slowly than those in gases.

Evidence for the movement of particles in a liquid was first observed when pollen grains on the surface of water were seen to move. The random motion of visible particles (pollen grains) caused by much smaller, invisible particles (water molecules) is called **Brownian motion**, after the botanist Robert Brown who first observed this phenomenon.

Quick Questions

1 Liquids and solids cannot be compressed as much as gases can. Explain why this should be the case.

2 Copy and complete the table below:

Material	Solid, liquid or gas	Property which makes it useful for the job
Wood for chairs		
Aluminium tubing for bicycles		
Petrol for cars		
Air in bicycle tyres		

3 Explain how the particles move in:

a) A solid;
b) A liquid;
c) A gas.

1.2 Gas laws

The pressure inside a balloon is caused by the gas particles striking the inside surface of the balloon. There is an increased pressure inside the balloon at higher temperatures as the gas particles have more energy and therefore move around faster, striking the inside surface of the balloon more frequently.

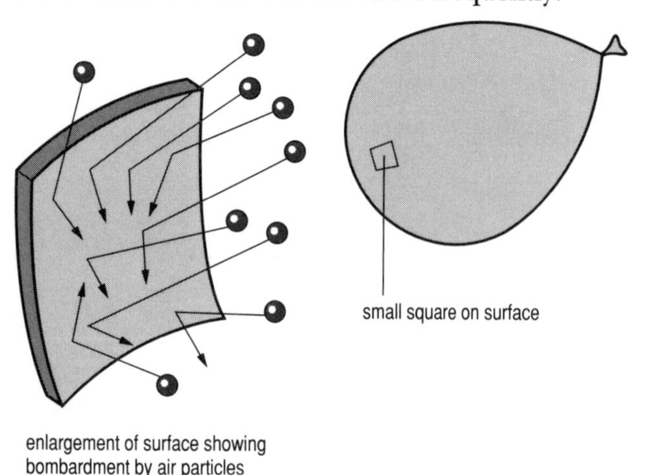

Figure 1.2.1 The gas particles striking the surface create the pressure

Since the balloon is an elastic envelope, the increased pressure causes the skin to stretch and the volume to increase. In 1781, a French scientist called J.A.C. Charles concluded that when a gas is kept in a vessel which allows its volume to change freely at constant pressure, i.e. like a syringe, its volume would increase when its temperature increases.

Investigating Charles' law

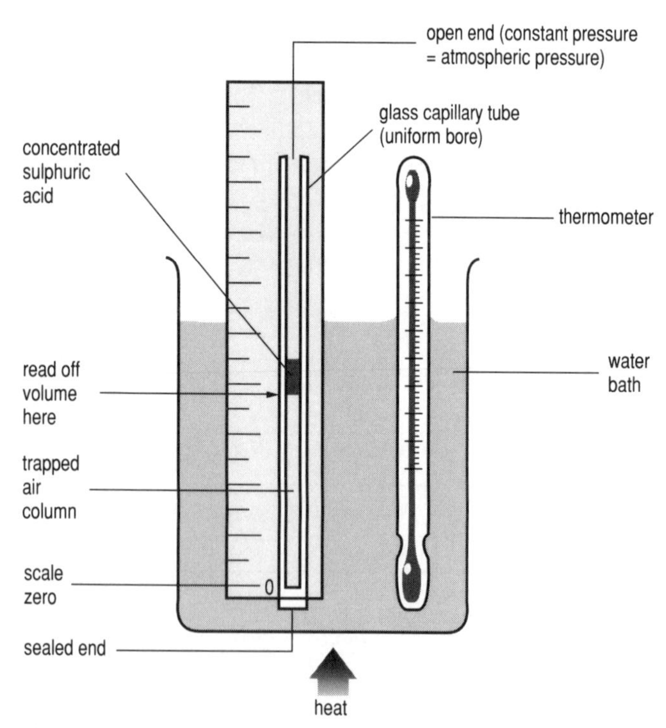

Figure 1.2.2 Charles' law apparatus

Charles' law applies to a fixed mass of gas. The air column in the glass tube shown in Figure 1.2.2 has been trapped by a drop of concentrated acid which moves up the tube as the gas expands.

The length of the air column can be taken as a measure of the volume of air that is trapped. Readings are taken of the temperature and the length of the air column as the water bath is heated.

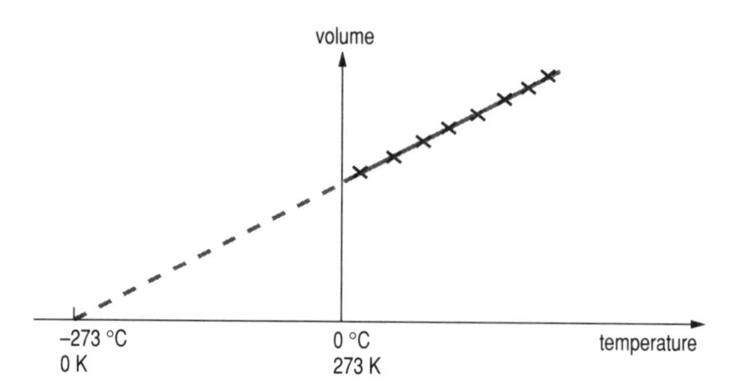

Figure 1.2.3 A graph of volume against temperature

The graph shows that at −273 °C, the volume of the gas should contract to zero! −273°C is called absolute zero and it is the lowest possible temperature. At this temperature, theoretically, particles have no motion and therefore possess no energy.

A new temperature scale was proposed by Lord Kelvin in 1854, called the Kelvin scale of temperature which has 0 K at absolute zero. Kelvins are the same size as degrees on the Celsius scale. On the Kelvin scale water freezes at 273 K and boils at 373 K. Note that we write 273 K without a '°' (degree) sign. In general, to convert a Celsius temperature to a Kelvin temperature, add 273.

$$K = °C + 273$$

Charles' law states the volume, V, of a fixed mass of gas is directly proportional to its absolute temperature, T, if the pressure is kept constant.

$$V \propto T$$

$$\text{or } V = \text{constant} \times T$$

$$\text{or } \frac{V}{T} = \text{constant}$$

Boyle's law

You can feel the increased pressure of the gas on your finger by pushing in the piston of a bicycle pump. When you do this, you push the same number of particles into a smaller volume (see Figure 1.2.4). This means they hit the walls of the pump more often, so increasing the pressure. In 1662 a scientist called Robert Boyle deduced that when the pressure was increased, the volume of the gases reduced.

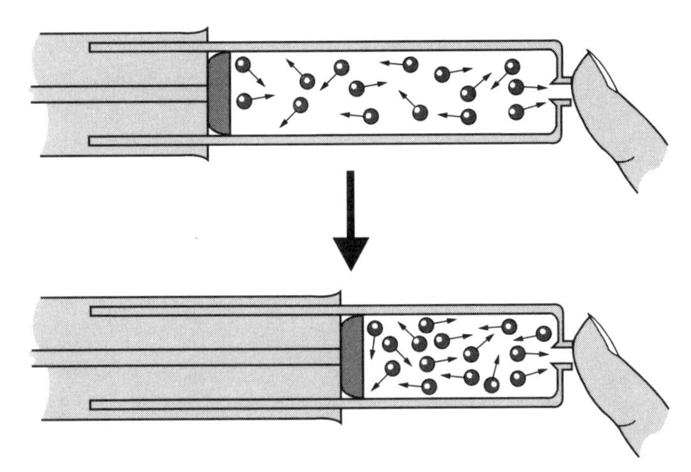

Figure 1.2.4 A higher pressure is created by pushing in the piston

Investigating Boyle's law

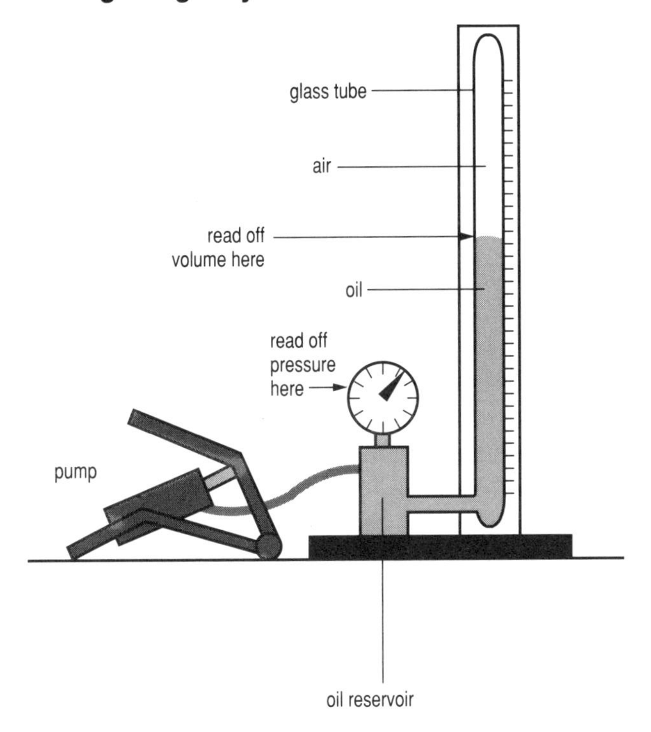

Figure 1.2.5 Boyle's law apparatus

The pressure on a fixed mass of air, trapped by oil in the strong glass tube, is increased. The length of the air column is measured at different pressures.

A graph plotted of volume against $1/p$ gives a straight line through the origin. This demonstrates **Boyle's law** which states that the volume, V, of a fixed mass of gas is inversely proportional to its pressure, p, if the temperature is constant. Mathematically,

$$V \propto \frac{1}{p}$$

$$\text{or } V = \frac{\text{constant}}{p}$$

$$\text{or } pV = \text{constant}$$

Combining the gas laws

Combining Boyle's and Charles' laws for a fixed mass of gas gives:

$$\frac{pV}{T} = \text{constant}$$

$$\text{or } \frac{p_1V_1}{T_1} = \frac{p_2V_2}{T_2}$$

where p_1, V_1 and T_1 and p_2, V_2 and T_2 are the pressure, volume and temperature (in Kelvin) in two different situations.

An example of the use of this relationship is given below: If the volume of a gas collected at 40 °C and 1×10^5 Pa pressure was 100 cm^3 what would be the volume of the gas at a temperature of 10 °C and a pressure of 2×10^5 Pa?

$$\text{Using } \frac{p_1V_1}{T_1} = \frac{p_2V_2}{T_2}$$

$p_1 = 1 \times 10^5$ Pa
$V_1 = 100$ cm^3
$T_1 = (40 + 273)$K
$= 313$ K

$p_2 = 2 \times 10^5$ Pa
$V_2 = ?$
$T_2 = (10 + 273)$K
$= 283$ K

$$\frac{(1 \times 10^5) \times 100}{313} = \frac{(2 \times 10^5) \times V_2}{283}$$

$$V_2 = \frac{(1 \times 10^5) \times 100 \times 283}{313 \times (2 \times 10^5)}$$

$$V_2 = 45.21 \text{ cm}^3$$

Quick Questions

1 When a gas is cooled, the particles move more slowly. Explain what will happen to the volume of the cooled gas if the pressure is kept constant.

2 A bubble of methane gas rises from the bottom of the North Sea.
   a) What will happen to the size of the bubble as it rises to the surface?
   b) Explain your answer to **a)**.

3 A gas syringe contains 50 cm^3 of oxygen gas at 20 °C. If the temperature was increased to 45 °C, what would be the volume occupied by this gas – assuming constant pressure throughout?

4 A bicycle pump contains 50 cm^3 of air at a pressure of 1×10^5 Pa. What would be the volume of the air if the pressure was increased to 2.1×10^5 Pa at constant temperature?

5 If the volume of a gas collected at 60 °C and 1×10^5 Pa pressure was 70 cm^3, what would be the volume at a temperature of 0 °C and a pressure of 4×10^5 Pa?

1.3 Changes of state

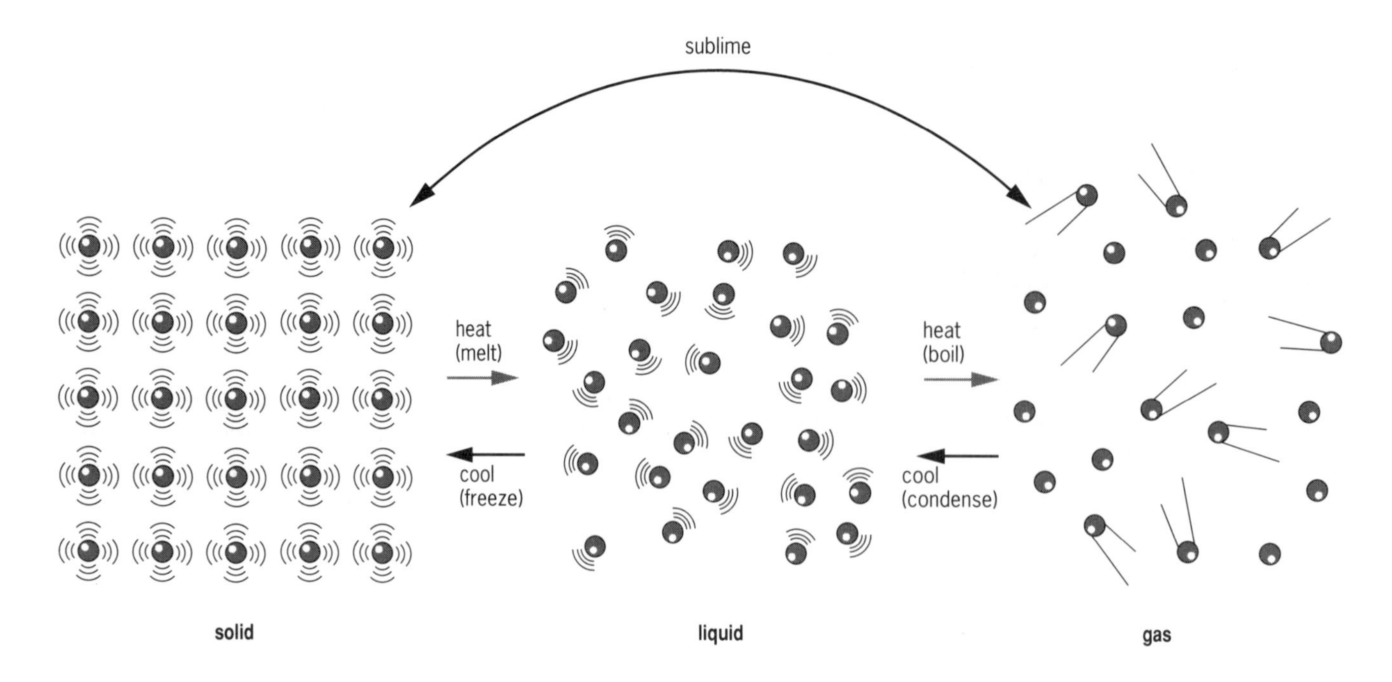

Figure 1.3.1 Summary of the changes of state

The kinetic theory model can be used to explain how a substance changes from one state to another.

If a solid is heated, the particles vibrate faster as they gain energy. This makes them 'push' their neighbouring particles further away from themselves. This causes an increase in the volume of the solid. Expansion takes place.

Eventually, the particles break free from one another as the energy from heating allows forces of attraction between the particles to be overcome and the bonds are broken. The regular pattern of the solid structure breaks down. The particles can now move around each other. The forces of attraction are now much less. The solid has melted. The temperature at which this takes place is called the **melting point** of the substance. The temperature of a pure melting solid will not rise until it has all melted. When the substance has become a liquid there are still some forces of attraction between the particles. This is why it is a liquid and not a gas.

Solids which have high melting points have stronger forces of attraction between their particles than those which have low melting points.

Some particles at the surface of the liquid have enough energy to overcome the forces of attraction between themselves and the other particles in the liquid and they escape and form a gas. If the liquid is heated, the particles will move around even faster as their average energy increases. The liquid begins to evaporate even faster.

Eventually, a temperature is reached at which the particles are trying to escape from the liquid so quickly that bubbles of gas actually start to form inside the bulk of the liquid. This temperature is called the **boiling point** of the substance.

Liquids with high boiling points have stronger forces between their particles than liquids with low boiling points.

When a gas is cooled, the average energy of the particles decreases and the particles move closer together. The attractive forces between the particles now cause the gas to **condense** into a liquid. When a liquid is cooled it freezes to form a solid. In each of these changes energy is given out.

The changes of state shown in Figure 1.3.1 are examples of physical changes. During a physical change no new substance is formed.

An unusual change of state

There are a few substances which change directly from a solid to a gas without even becoming a liquid when they are heated. This rapid spreading out of the particles is called sublimation. Cooling causes a change from a gas directly back to a solid. Examples of substances which behave in this way are carbon dioxide and iodine.

In the case of carbon dioxide, at temperatures below −78 °C it is a white solid called dry ice. When heated just above −78 °C, it changes into carbon dioxide gas.

Heating and cooling curves

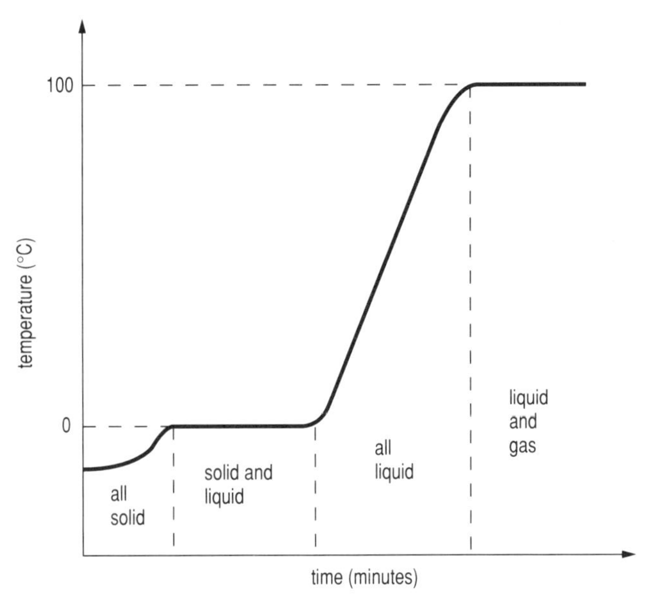

Figure 1.3.2 Graph of temperature against time for the change from ice to water to steam

You can see from the curve that changes of state have taken place. When the temperature was first measured, only ice was present. After two minutes the curve flattens, showing that even though heat energy is being put in, the temperature remains constant.

In ice, the particles of water are close together and are attracted to one another. For ice to melt, the particles must obtain sufficient energy to overcome the forces of attraction between the water particles. This is where the heat energy is going.

The temperature will only begin to rise again after all the ice has melted. Generally, the heating curve for a pure solid always stops rising at its melting point and gives rise to a sharp melting point.

In the same way, if you want to boil a liquid such as water you have to give it some extra energy. This can be seen on the graph where the curve levels out at the boiling point of water.

The reverse processes of condensing and freezing occur on cooling. Then, however, energy is given out when the gas condenses to the liquid, and the liquid freezes to give the solid.

Kinds of particle

There are three kinds of **particle - atoms**, **molecules** and **ions**. Atoms are the simplest of these. An atom is the smallest part of an element. For example, hydrogen is made up of hydrogen atoms and oxygen is made up of oxygen atoms. Sometimes, atoms join together in groups called molecules. For example, hydrogen and oxygen atoms combine to form water molecules.

Atoms are uncharged particles. An ion is a charged particle as it carries an electrical charge. For example, a hydrogen ion has a positive charge and a hydroxide ion has a negative charge (see page 000). Most ions are formed when atoms undergo changes during chemical reactions.

Atoms, molecules and ions are all very small. For example:

- A hydrogen atom is 1×10^{-10} m (0.0000000001 m) across. It is the smallest atom.
- Glucose is quite a large molecule containing hydrogen, carbon and oxygen. But even this molecule is only 1×10^{-9} m across.

```
H    O
 \  //
   C
   |
H—C—OH
   |
HO—C—H
   |
H—C—OH
   |
H—C—OH
   |
   C
 / | \
H  |  OH
   H
```

Figure 1.3.3 The formula of glucose is $C_6H_{12}O_6$

Quick Questions

1. Draw and label a graph you would expect to produce if water at 100 °C was allowed to cool to a temperature of −15 °C.
2. Explain why bromine would diffuse faster in evacuated rather than unevacuated gas jars.
3. Explain why diffusion in liquids is slower than in gases.
4. Use the kinetic theory to explain the following:
   a) You can smell hot coffee several metres away, yet you have to be near cold coffee to smell it.
   b) When a person wearing perfume enters the room it takes several minutes for the smell to reach the back of the room.
5. What is meant by the terms:
   a) Atom;
   b) Molecule;
   c) Ion?
   Use an example to help with each explanation.

1.4 Elements, compounds and mixtures

Robert Boyle suggested the name **element** for any substance that cannot be broken down into a simpler substance. Each element is made up of only one kind of atom. The word atom comes from the Greek word for 'unsplittable'.

For example, aluminium is an element which is made up of only aluminium atoms. It is not possible to obtain a simpler substance chemically from the aluminium atoms. You can only make more complicated substances from it, such as aluminium oxide, aluminium nitrate or aluminium sulphate.

There are 112 elements which have now been identified. 92 of the elements occur naturally. They range from some very reactive gases, such as fluorine, to gold and platinum, which are unreactive elements.

All these different elements can be classified according to their various properties. Table 1.4.1 below shows the physical data for some common metallic and non-metallic elements.

Table 1.4.2 gives a summary of the different properties of metals and non-metals.

Table 1.4.1 Physical data for some metallic and non-metallic elements at room temperature and pressure

Element	Metal/non-metal	Density ($g\ cm^{-3}$)	Melting point (°C)	Boiling point (°C)
Aluminium	Metal	2.70	660	2580
Iron	Metal	7.87	1535	2750
Copper	Metal	8.92	1083	2567
Gold	Metal	19.29	1065	2807
Carbon	Non-metal	2.25	2652	Sublimes
Oxygen	Non-metal	1.15[a]	−218	−183
Sulphur	Non-metal	2.07	113	445
Nitrogen	Non-metal	0.88[b]	−210	−196

Source: Earl B., Wilford L.D.R., Chemistry Data Book. Nelson Blackie, 1991

[a] At −184 °C

[b] At −197 °C

Table 1.4.2 Different properties of metals and non-metals

Property	Metal	Non-metal
Physical state at room temperature	Usually solid (occasionally liquid)	Solid, liquid or gas
Malleability	Good	No – usually soft or brittle
Ductility	Yes	Very poor
Appearance (solids)	Shiny (lustrous)	Dull
Melting point	Usually high	Usually low
Boiling point	Usually high	Usually low
Density	Usually high	Usually low
Conductivity (thermal and electrical)	Good	Very poor

Mixtures and compounds

Many everyday things are not pure substances, they are **mixtures** of elements and/or **compounds**. The differences between compounds and mixtures can be demonstrated by considering the reaction between iron filings and sulphur. A mixture of iron filings and sulphur looks different from the individual elements but has the properties of both iron and sulphur. For example, a magnet can be used to separate the iron filings from the sulphur.

Substances in a mixture have not undergone a chemical reaction. If the mixture of iron and sulphur is heated, a new substance is formed called iron(II) sulphide. The word equation for this reaction is:

$$\text{iron} + \text{sulphur} \xrightarrow{\text{heat}} \text{iron(II) sulphide}$$

During the reaction, heat energy is given out as new chemical bonds are formed. This is called an exothermic reaction and accompanies a **chemical change**. This type of change is one in which a permanent change has taken place.

The iron(II) sulphide formed has totally different properties to the mixture of iron and sulphur - it is a compound. Table 1.4.3 summarises some of the differences.

In iron(II) sulphide, FeS, one atom of iron has combined with one atom of sulphur. Because the atoms have not chemically combined, the mixture of iron and sulphur has no such ratio. Table 1.4.4 summarises the major differences between mixtures and compounds.

Table 1.4.3 Different properties of iron, sulphur, an iron/sulphur mixture and iron(II) sulphide

Substance	Appearance	Effect of a magnet	Effect of dilute hydrochloric acid
Iron	Dark grey powder	Attracted to it	Very little action when cold. When warm, a gas is produced with a lot of bubbling (effervescence)
Sulphur	Yellow powder	None	No effect when hot or cold
Iron/sulphur mixture	Dirty yellow powder	Iron powder attracted to it	Iron powder reacts as above
Iron(II) sulphide	Dark grey solid	No effect	A foul-smelling gas is produced with some effervescence

Table 1.4.4 The major differences between mixtures and compounds

Mixture	Compound
It contains two or more substances	It is a single substance
The composition can vary	The composition is always the same
No chemical change takes place when a mixture is formed	When the new substance is formed it involves chemical change
The properties are those of the individual elements	The properties are very different to those of the component elements
The components may be separated quite easily by physical means	The components can only be separated by one or more chemical reactions

Composite materials

Composite materials are those which combine the properties of two constituents, in order to get the exact properties needed for a particular job.

Glass-reinforced plastic (GRP) is an example of a composite material. It is made by embedding short fibres of glass in a matrix of plastic. The glass fibres give the plastic extra strength so that it does not break when it is bent or moulded into shape. The finished product has the lightness of plastic, as well as the strength and flexibility of the glass fibres. GRP is used in the manufacture of boats and car bodies. Other examples of composite material are bone (protein and calcium phosphate) and wood (cellulose fibres and lignin).

Quick Questions

1 Helium, oil, air, mercury, water, copper, milk, carbon monoxide, salt, solder.
Which of the above is:
a) A liquid compound;
b) An alloy;
c) A solid compound;
d) A mixture?

2 Why are composite materials often used instead of single materials?

1.5 Separating mixtures

Many mixtures contain useful substances mixed with unwanted material. In order to obtain these useful substances, chemists often have to separate them from the impurities.

Separating solid/liquid mixtures

If a solid substance is added to a liquid, it may dissolve to form a **solution**. In this case the solid is said to be **soluble** and is called the **solute**. The liquid it has dissolved in is called the **solvent**. An example of this type of process is when sugar is added to tea or coffee.

Sometimes a solid does not dissolve in a given liquid. The solid is said to be insoluble in that liquid. For example, when making tea the solid tea leaves do not dissolve in boiling water when tea is made from them.

Filtering

Filtration is a common separation technique used in chemistry laboratories throughout the world. It is used when a solid needs to be separated from a liquid. For example, sand can be separated from its mixture with water by filtering through the filter paper as shown in the diagram below.

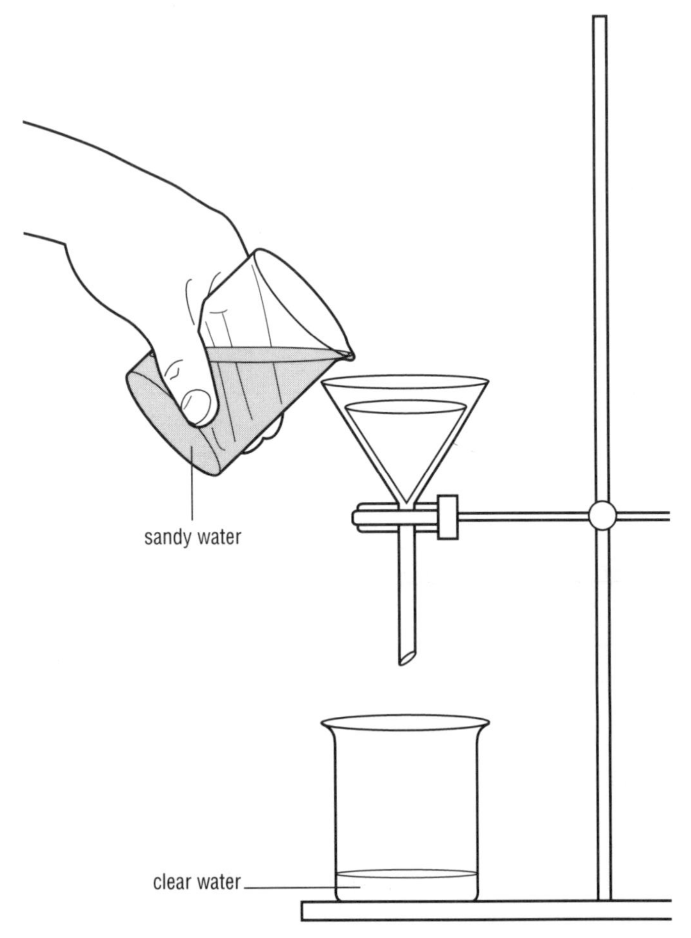

Figure 1.5.1 Filtering sand from the sand–water mixture

The filter paper contains holes which are small enough to allow the molecules of water through, but not the sand particles. It acts like a sieve. The sand gets trapped in the filter paper and the water passes through it. The sand is called the **residue** whilst the water is called the **filtrate**.

Decanting

Carrots do not dissolve in water. When you have boiled some carrots it is easy to separate them from the water by pouring it off. This process is called decanting. This technique is used quite often to separate an **insoluble** solid, which has settled to the bottom of a flask, from a liquid.

Crystallisation

In many parts of the world, salt is obtained from sea water on a vast scale. This is done by using the heat of the sun to evaporate the water, leaving a saturated solution of salt known as brine. A **saturated solution** is defined as one that contains as much solute as can be dissolved at a particular temperature. When the solution is saturated, the salt begins to crystallise and it is removed using large scoops.

Centrifuging

Another way to separate a solid from a liquid is to use a centrifuge. This technique is sometimes used instead of filtration. It is usually used when the solid particles are so small that they spread out (disperse) throughout the liquid and remain in **suspension**. They do not settle to the bottom of the container because they are light enough to be held in suspension by collision and attraction with solvent particles. The technique of **centrifuging** or centrifugation involves the suspension being spun round very fast in a centrifuge so that the solid gets flung to the bottom of the tube.

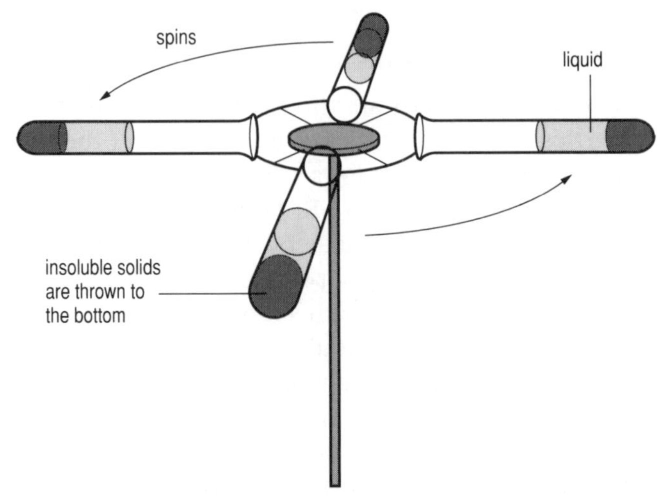

Figure 1.5.2 The sample is spun around very fast and the solid is flung to the bottom of the tube

After the solid has been forced to the bottom of the tube, the pure liquid can be decanted off. This method of separation is used extensively to separate blood cells from liquid blood plasma.

Evaporation

If the solid has dissolved in the liquid it cannot be separated by filtering or centrifuging. Instead, the solution can be heated so that the liquid evaporates completely, leaving the solid behind. The simplest way to obtain salt from its solution is by slow evaporation, as shown in Figure 1.5.3. This process is used to produce crystals from solutions - the process is called **crystallisation**.

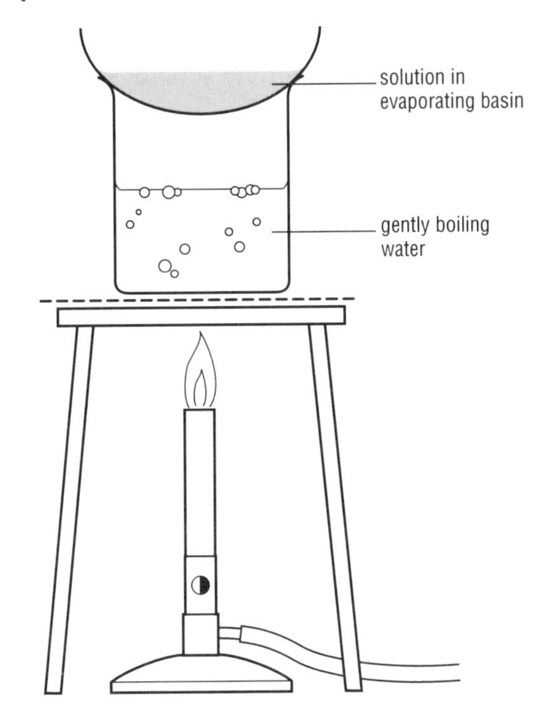

Figure 1.5.3 Apparatus used to slowly evaporate a solvent

Simple distillation

If we want to obtain the solvent from a solution then the process of **distillation** can be carried out. The apparatus used in this process is shown in Figure 1.5.4.

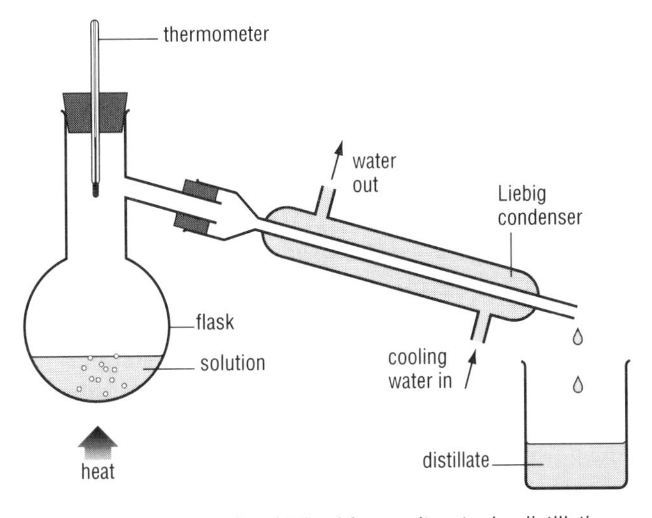

Figure 1.5.4 Water can be obtained from salt water by distillation

Water can be obtained from salt water using this method. The solution is heated in the flask until it boils. The steam rises into the Liebig condenser where it condenses back into water. The salt is left behind in the flask. In hot and arid countries such as Saudia Arabia, this sort of technique is used on a much larger scale to obtain pure water for drinking.

Separating liquid/liquid mixtures

In recent years there have been many oil tanker disasters that have resulted in millions of litres of oil being washed into the sea. Oil and water do not easily mix. They are said to be **immiscible**. When cleaning up disasters of this type, a range of chemicals can be added to the oil which makes it more soluble. This results in the oil and water mixing with each other. They are now said to be **miscible**. The following techniques can be used to separate mixtures of liquids.

Liquids which are immiscible

If two liquids are immiscible they can be separated using a separating funnel. The mixture is poured into the funnel and the layers allowed to separate. The lower layer can then be run off by opening the tap as shown in Figure 1.5.5.

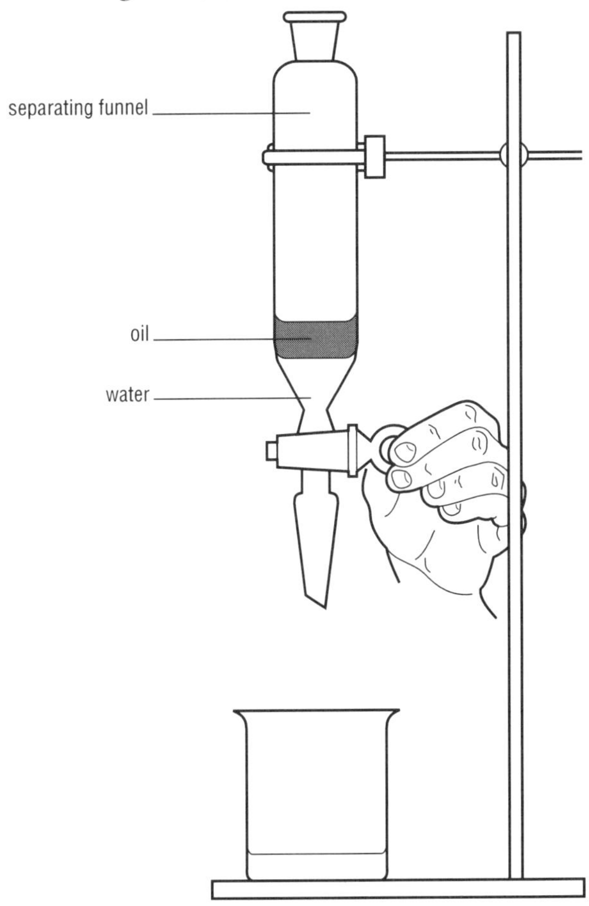

Figure 1.5.5 The water is more dense than the oil, so it sinks to the bottom of the separating funnel. When the tap is opened the water can be run off

Liquids which are miscible

If miscible liquids are to be separated then it can be carried out by **fractional distillation**. The apparatus used for this process is shown in Figure 1.5.6.

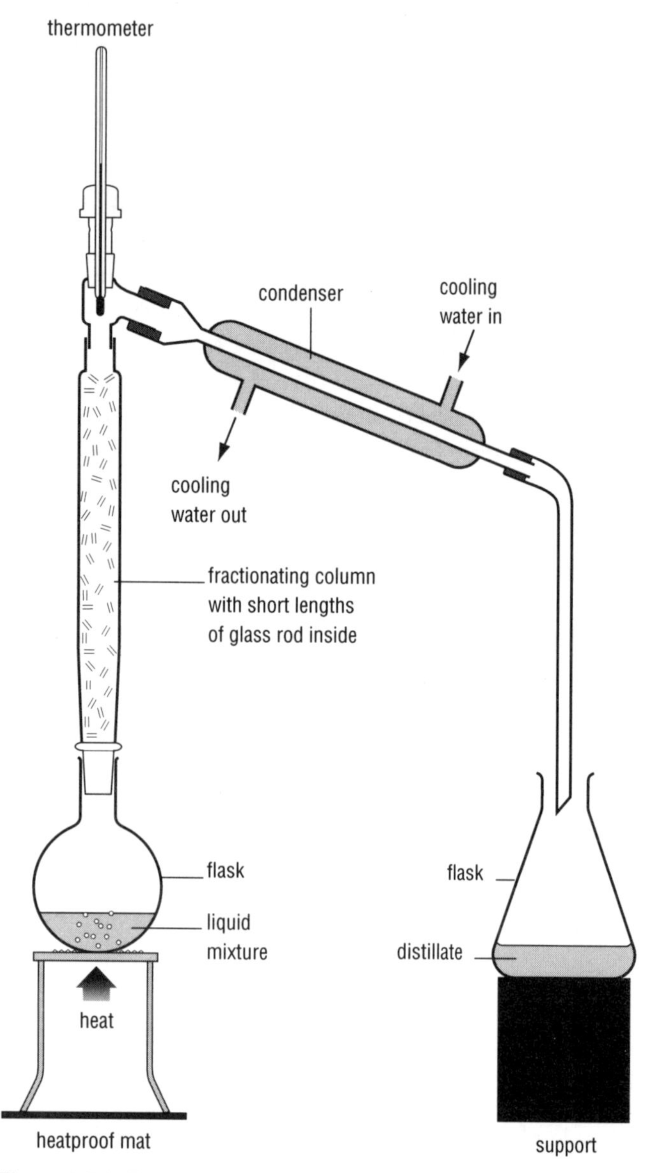

Figure 1.5.6 Typical fractional distillation apparatus

The apparatus shown could be used to separate a mixture of ethanol and water. The process relies upon the liquids having different boiling points.

Ethanol boils at 78 °C while water boils at 100 °C. When the mixture is heated the vapour produced is mainly ethanol with some steam. Because water has the higher boiling point it condenses out from the mixture with ethanol. This is what takes place in the fractionating column. The water condenses and drips back into the flask but the ethanol vapour moves up the column and into the condenser. Here it condenses into liquid ethanol and is collected in the receiver.

The technique is used extensively to separate the individual gases from the air as well as to separate **crude oil** into its constituent substances.

Separating solid/solid mixtures

When iron is separated from sulphur using a magnet, we use one of the properties of iron, that is, it is **magnetic**. In a similar way, it is possible to separate scrap iron from other metals by using a large electromagnet.

It is, therefore, essential that when separating solid/solid mixtures that you pay particular attention to the individual physical properties of the components. If, for example, you wish to separate two solids, one of which sublimes, then use this property in the method you employ.

In the case of an iodine/salt mixture the iodine sublimes but salt does not. Iodine can be separated by heating the mixture in a fume cupboard as shown in Figure 1.5.7. The iodine sublimes and reforms on the cool inverted funnel.

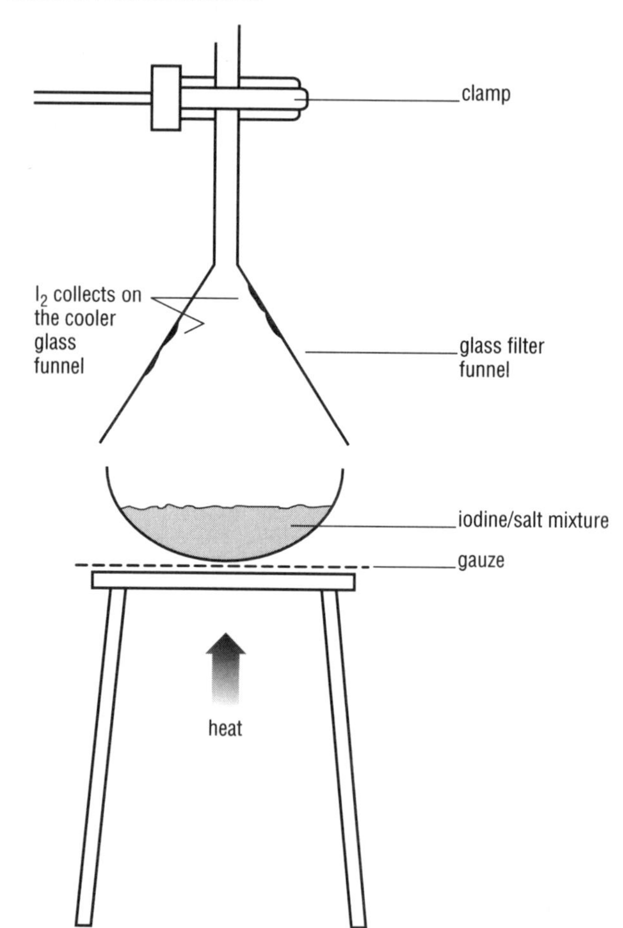

Figure 1.5.7

Chromatography

Chromatography can be used to separate two or more solids which are soluble in the same solvent.

There are several types of chromatography; however they all follow the same basic principles. The simplest kind is called paper chromatography. To separate the different coloured dyes in a sample of black ink, a spot of the ink is put onto a piece of chromatography paper. This paper is then set in a suitable solvent as shown in Figure 1.5.8.

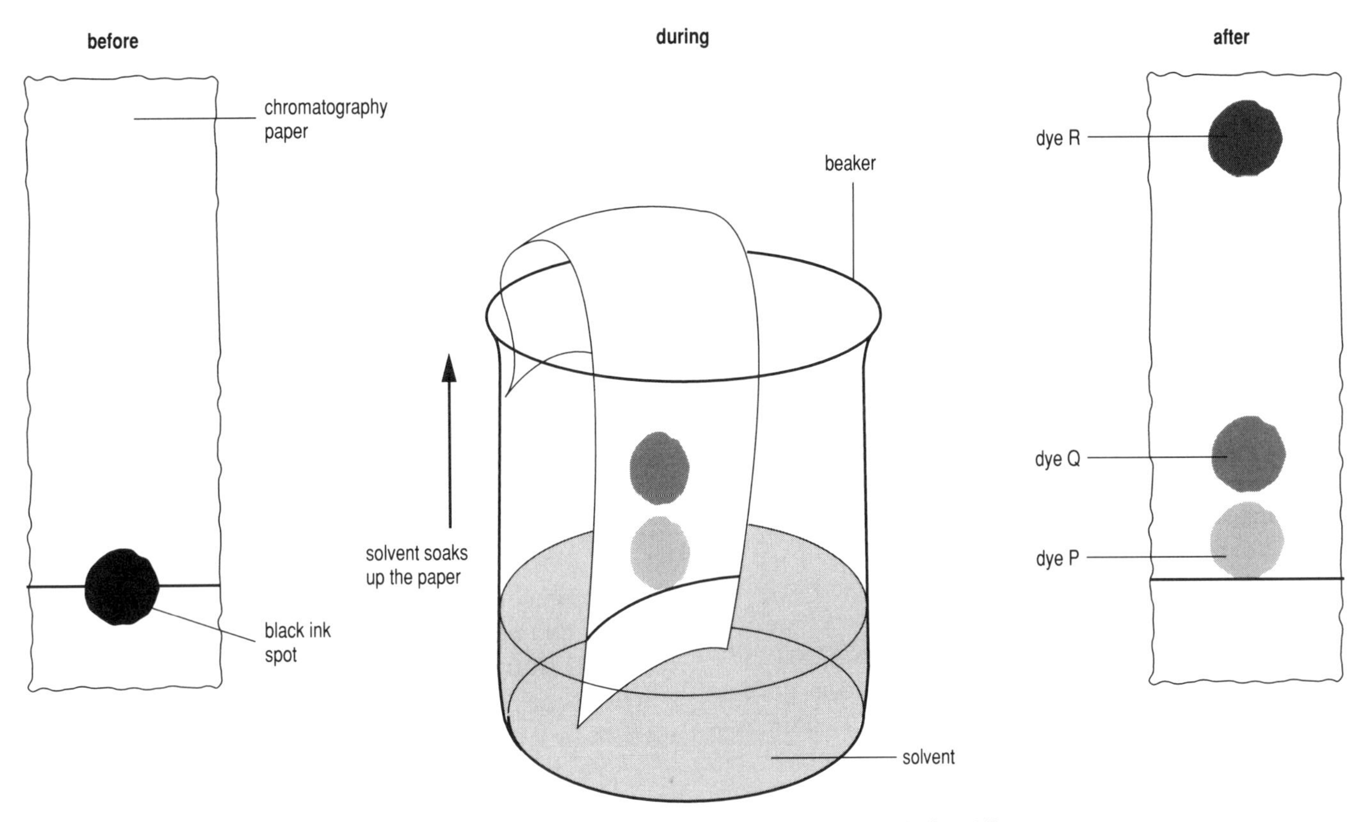

Figure 1.5.8 Chromatographic separation of black ink. The black ink separates into three dyes: P, Q and R

As the solvent moves up the paper the dyes are carried with it and begin to separate. They separate because the substances have different solubilities in the solvent and are absorbed to different degrees by the chromatography paper. The chromatogram in Figure 1.5.8 shows how the ink can be separated into the three dyes P, Q and R.

Chromatography is used extensively in hospital, health and forensic science laboratories to separate a variety of mixtures.

The substances to be separated do not need to be coloured. Colourless substances can be made to show up by spraying the chromatogram with a locating agent. The locating agent will react with the colourless substances to form a coloured product. In other situations the position of the substances on the chromatogram may be located using ultra violet light.

Unlike other separation techniques, chromatography is not (usually) used for obtaining actual samples of the parts of the mixture. Usually the technique indicates that the parts are there and what they are.

Solvent extraction

Sugar can be obtained from crushed sugar cane by adding water. The water dissolves the sugar from the sugar cane. This is an example of solvent extraction. In a similar way, some of the green substances can be removed from ground-up grass using ethanol. The substances are extracted from a mixture by using a solvent which dissolves only those substances required.

Quick Questions

1. Write down as many examples as you can think of in which a centrifuge is used.
2. What is the difference between fractional distillation and simple distillation?
3. Describe how you would use chromatography to show that grass contains a mixture of green pigments.
4. Explain the following terms:
   - **a)** Miscible;
   - **b)** Immiscible;
   - **c)** Evaporation;
   - **d)** Condensation;
   - **e)** Solvent extraction.
5. Devise a method for obtaining salt from sea water in the school laboratory.
6. Name the method which is most suitable for separating the following:
   - **a)** Ethanol (boiling point 78 °C) from a mixture of ethanol and water;
   - **b)** The sediment formed at the bottom of a wine bottle;
   - **c)** Nitrogen from liquid air;
   - **d)** Red blood cells from plasma;
   - **e)** Petrol and kerosine from crude oil;
   - **f)** Coffee grains from coffee solution;
   - **g)** Pieces of steel from engine oil;
   - **h)** A complex mixture of proteins.

1.6 What's in an atom?

Until just over 100 years ago scientists believed that atoms were solid particles like marbles. However, in the last hundred years it has been proved that atoms are in fact made up of even smaller **sub-atomic** particles. The most important of these are **electrons**, **protons** and **neutrons**.

Inside atoms

These three sub-atomic particles are found in two distinct and separate regions. The protons and neutrons are found in the centre of the atom which is called the **nucleus**. The neutrons have no charge but protons are positively charged.

The nucleus occupies only a very small volume of the atom but it is very dense.

Around the nucleus is the rest of the atom, where electrons are likely to be found. The electrons are **negatively charged** and move around very quickly in **electron shells** or **electron energy levels**.

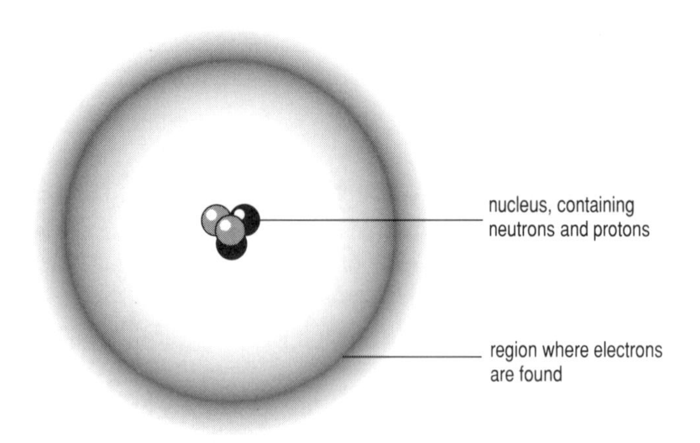

Figure 1.6.1 Diagram of an atom

The electrons are held within the atom by an **electrostatic force of attraction** between themselves and the positive charge of the protons in the nucleus. A summary of the particles is shown in Table 1.6.1.

Table 1.6.1 Characteristics of a proton, a neutron and an electron

Particle	Symbol	Relative mass (amu)	Relative charge
Proton	p	1	+1
Neutron	n	1	0
Electron	e	1/1837	−1

The masses are measured in atomic mass units, because they are so light that they cannot be measured accurately in grams.

Although atoms contain electrically charged particles the atoms themselves are electrically neutral. This is because atoms contain equal numbers of electrons and protons.

Atomic number and mass number

The number of protons in the nucleus of an atom is called the **atomic number** (or **proton number**) and is given the symbol **z**. Each element has its own atomic number and no two elements have the same atomic number.

Electrons possess very little mass. So the mass of any atom depends mainly on the number of neutrons and protons in its nucleus. The total number of protons and neutrons is called the **mass number** (or **nucleon number**) and is given the symbol **A**.

The atomic number and mass number of an element are usually written in the following shorthand way:

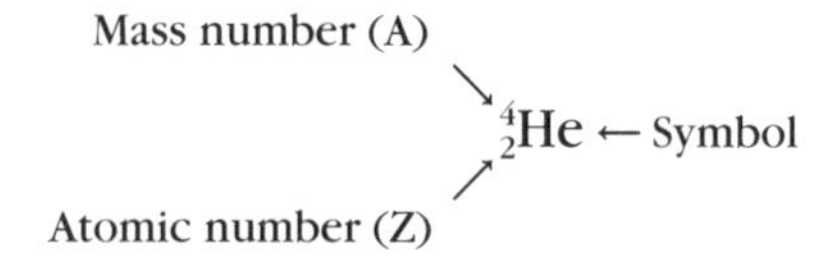

The number of neutrons present can be calculated by rearranging the relationship between the atomic number, mass number and number of neutrons to give:

number of neutrons = mass number (A) − atomic number (Z)

For example, the number of neutrons in one atom of helium is:

$$4 - 2 = 2$$

Table 1.6.2, on the next page, shows the number of protons, neutrons and electrons in the atoms of some common elements.

Isotopes

Not all atoms in a sample of chlorine, for example, will be identical. Some atoms of the same element can contain different numbers of neutrons. Atoms of the same element which have different numbers of neutrons are called **isotopes**.

Generally isotopes behave in the same way during chemical reactions. The only effect of the extra neutron is to alter the mass of the atom and properties which depend on it, such as density.

Some of the atoms of certain isotopes are unstable because of the extra number of neutrons and are said to be **radioactive**. The best known elements which have radioactive isotopes are uranium and carbon.

Table 1.6.2 Number of protons, neutrons and electrons in some elements

Element	Symbol	Atomic number	Number of electrons	Number of protons	Number of neutrons	Mass number
Hydrogen	H	1	1	1	0	1
Helium	He	2	2	2	2	4
Carbon	C	6	6	6	6	12
Nitrogen	N	7	7	7	7	14
Oxygen	O	8	8	8	8	16
Fluorine	F	9	9	9	10	19
Neon	Ne	10	10	10	10	20
Sodium	Na	11	11	11	12	23
Magnesium	Mg	12	12	12	12	24
Sulphur	S	16	16	16	16	32
Potassium	K	19	19	19	20	39
Calcium	Ca	20	20	20	20	40
Iron	Fe	26	26	26	30	56
Copper	Cu	29	29	29	35	64

Table 1.6.3 Isotopes of carbon and uranium

Element	${}^{A}_{Z}$Symbol	Particles present
Carbon	${}^{12}_{6}C$ ${}^{13}_{6}C$ ${}^{14}_{6}C$	6p, 6e, 6n 6p, 6e, 7n 6p, 6e, 8n
Uranium	${}^{235}_{92}U$ ${}^{238}_{92}U$	92p, 92e, 143n 92p, 92e, 146n

Relative atomic mass

The average mass of a large number of atoms of an element is called its relative atomic mass, **RAM**, symbol $\mathbf{A_r}$. This quantity takes into account the percentage abundance of all the isotopes of an element which exist.

Table 1.6.4

	${}^{35}_{17}Cl$	${}^{37}_{17}Cl$
Particles present	17p,17e, 18n	17p,17e, 20n
Atomic Mass	35	37
% Abundance	75	25

The RAM of an element is defined as the average mass of its isotopes compared to one twelfth the mass of one atom of carbon-12.

$$A_r = \frac{\text{average mass of isotopes of the element}}{1/12 \times \text{mass of 1 atom of carbon 12}}$$

The 'average mass' of a chlorine atom is:

$$\text{Average mass (RAM)} = \frac{(75 \times 35) + (25 \times 37)}{100} = 35.5$$

$$\text{Hence, } A_r = \frac{35.5}{1} = 35.5$$

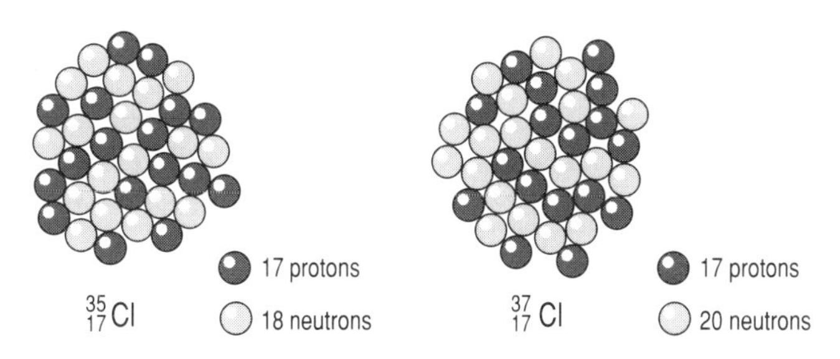

Figure 1.6.2 The two isotopes of chlorine

Quick Questions

1. What is meant by the terms proton number and nucleon number?
2. Name each of the following atoms and write down how many protons, electrons and neutrons they have:
   a) ${}^{12}_{6}C$ **b)** ${}^{16}_{8}O$ **c)** ${}^{27}_{13}Al$ **d)** ${}^{31}_{15}P$ **e)** ${}^{64}_{29}Cu$
3. **a)** ${}^{238}U$ and ${}^{235}U$ are isotopes of uranium. With reference to this example, explain what you understand by the term isotope.
   b) Given that the percentage abundance of ${}^{20}Ne$ is 90% and that of ${}^{22}Ne$ is 10%, calculate the A_r of neon.

1.7 The arrangement of electrons

The electrons, the lightest of the sub-atomic particles, move quickly around the nucleus, in electron shells or energy levels rather like the planets orbiting around the Sun – Figure 1.7.1.

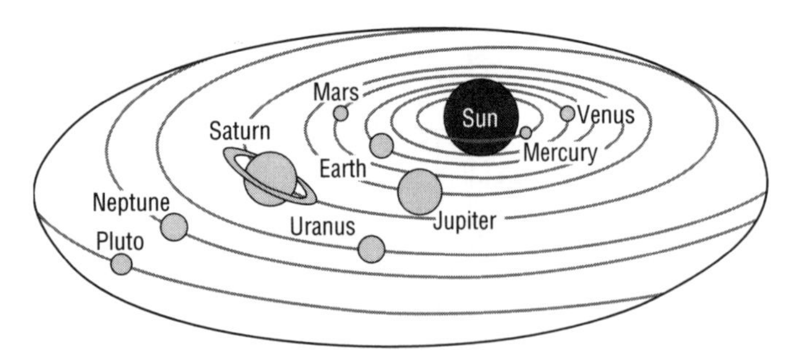

Figure 1.7.1 The way electrons are arranged in an atom is similar to the way the planets are in orbit around the Sun

Electrons can only occupy certain, definite energy levels and cannot exist between them. Each of the electron energy levels can only hold a certain number of electrons.

1st energy level holds up to 2 electrons.
2nd energy level holds up to 8 electrons.
3rd energy level holds up to 8 electrons (but in certain circumstances can hold up to 18).

There are further energy levels which contain increasing numbers of electrons.

The electrons fill the energy levels starting from the energy level nearest the nucleus, which has the lowest energy. When this is full (with two electrons) the next electron goes into the second energy level. When this energy level is full with eight electrons then the electrons begin to fill the third and fourth energy levels.

For example, the carbon atom has an atomic number of 6, and therefore it has six electrons. Two of the six electrons enter the first energy level leaving four to occupy the second energy level, as shown in Figure 1.7.2.

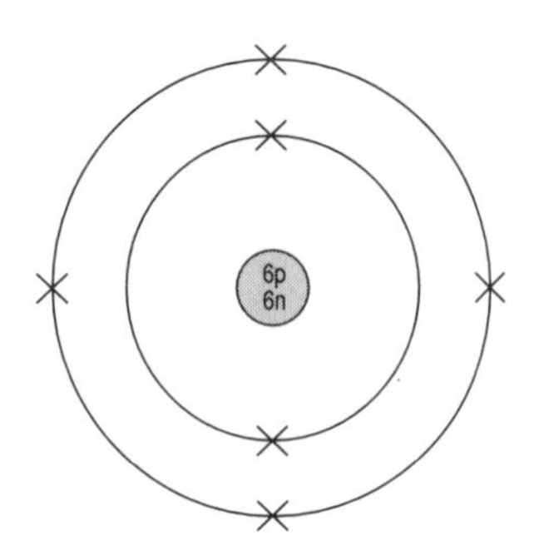

Figure 1.7.2 The arrangement of electrons in a carbon atom

The **electron configuration** (or **structure**) for carbon can be written in a shorthand way as 2.4.

There are 112 elements and the Table 1.7.1 shows the way in which the electrons are arranged in the first twenty of these elements.

Table 1.7.1 Electron arrangement in the first 20 elements

Element	Symbol	Atomic number	Number of electrons	Electron structure
Hydrogen	H	1	1	1
Helium	He	2	2	2
Lithium	Li	3	3	2,1
Beryllium	Be	4	4	2,2
Boron	B	5	5	2,3
Carbon	C	6	6	2,4
Nitrogen	N	7	7	2,5
Oxygen	O	8	8	2,6
Fluorine	F	9	9	2,7
Neon	Ne	10	10	2,8
Sodium	Na	11	11	2,8,1
Magnesium	Mg	12	12	2,8,2
Aluminium	Al	13	13	2,8,3
Silicon	Si	14	14	2,8,4
Phosphorus	P	15	15	2,8,5
Sulphur	S	16	16	2,8,6
Chlorine	Cl	17	17	2,8,7
Argon	Ar	18	18	2,8,8
Potassium	K	19	19	2,8,8,1
Calcium	Ca	20	20	2,8,8,2

Atoms with full outer energy levels are particularly stable and show little chemical reactivity. These elements are known as the noble gases or inert gases (see Section 2.1).

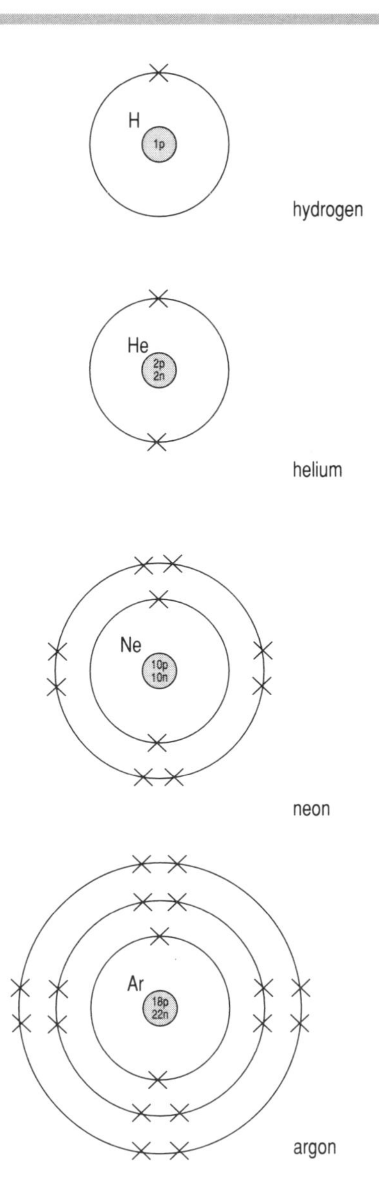

Figure 1.7.3 Electron arrangements of hydrogen, helium, neon and argon

Ions

An ion is an electrically charged particle. When an atom loses one or more electrons it becomes a positively charged ion. For example, during the chemical reactions of potassium, each atom loses an electron to form a positive ion, K^+.

$$_{19}K^+$$

$$\begin{aligned} 19 \text{ protons} &= 19+ \\ 18 \text{ electrons} &= 18- \\ \text{overall charge} &= 1+ \end{aligned}$$

When an atom gains one or more electrons it becomes a negatively charged ion. For example, during some of the chemical reactions of chlorine it gains an electron to form a negative ion, Cl^-.

$$_{17}Cl^-$$

$$\begin{aligned} 17 \text{ protons} &= 17+ \\ 18 \text{ electrons} &= 18- \\ \text{overall charge} &= 1- \end{aligned}$$

Table 1.7.2 shows some common ions. You will notice that:

- Some ions contain more than one type of atom, for example, NO_3^-.
- An ion may possess more than one unit of charge (either negative or positive), for example, Al^{3+}, O^{2-} and SO_4^{2-}.

Table 1.7.2 Some common ions

Name	Formula
Lithium ion	Li^+
Sodium ion	Na^+
Potassium ion	K^+
Magnesium ion	Mg^{2+}
Aluminium ion	Al^{3+}
Zinc ion	Zn^{2+}
Ammonium ion	NH_4^+
Fluoride ion	F^-
Chloride ion	Cl^-
Bromide ion	Br^-
Oxide ion	O^{2-}
Sulphide ion	S^{2-}
Nitrate ion	NO_3^-
Sulphate ion	SO_4^{2-}

Quick Questions

1 How many electrons may be accommodated in the first three energy levels?

2 Draw energy level diagrams for the following elements:
  a) Carbon;
  b) Sulphur;
  c) Calcium.

3 In what ways are the electron structures of carbon and silicon similar?

4 From the first 20 elements name or give the symbols of those elements:
  a) With three electrons in their outer energy level;
  b) Which have full outer energy levels.

5 An atom **X** has an atomic number of 19 and relative atomic mass of of 39.
  a) How many electrons, protons and neutrons are there in an atom of **X**?
  b) How many electrons will there be in the outer energy level (shell) of an atom of **X**?

1.8 Ionic bonding

Ionic bonds are usually found in compounds that contain metals combined with non-metals. When this type of bond is formed, electrons are transferred from the metal atoms to the non-metal atoms during the chemical reaction. In doing this, the atoms become more stable by getting full outer energy levels like the nearest inert gas.

For example, consider what happens when sodium and chlorine combine to make sodium chloride which contains sodium ions (Na^+) and chloride ions (Cl^-).

sodium atom	+	chlorine atom	$\longrightarrow$	sodium chloride
Na	+	Cl	$\longrightarrow$	Na^+Cl^-

Only the outer electrons are important in bonding, so we can simplify any diagram by missing out the inner energy levels (see Figure 1.8.1).

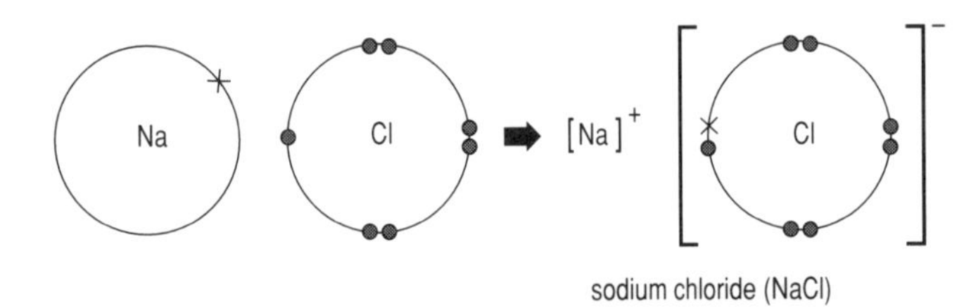

Figure 1.8.1 Simplified diagram of ionic bonding in sodium chloride

The charges on the sodium and chloride ions are equal but opposite. These oppositely charged ions attract each other and are pulled, or bonded, to one another by strong electrostatic forces. This type of bonding is called ionic bonding. The alternative name, electrovalent bonding, is derived from the fact that there are electric charges on the atoms involved in the bonding.

Figure 1.8.2 shows the electron transfers that take place during the formation of magnesium oxide.

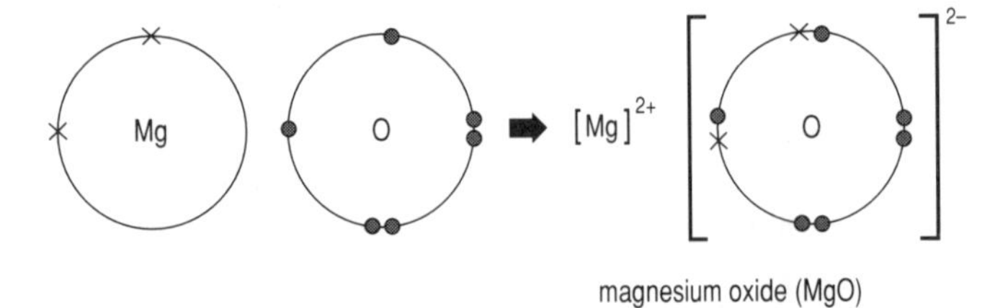

Figure 1.8.2 Simplified diagram of ionic bonding in magnesium oxide

Ionic structures

In ionic compounds the ions are packed together in a regular arrangement called a **lattice**.

Figure 1.8.3 shows only a tiny part of a small crystal of sodium chloride whose structure was determined from X-ray photographs. Many millions of sodium ions and chloride ions would be arranged in this way in a crystal of sodium chloride to make up the **giant ionic structure**. Within the lattice, oppositely charged ions attract one another strongly in all directions. Each sodium ion in the lattice is surrounded by six chloride ions, and each chloride ion is surrounded by six sodium ions.

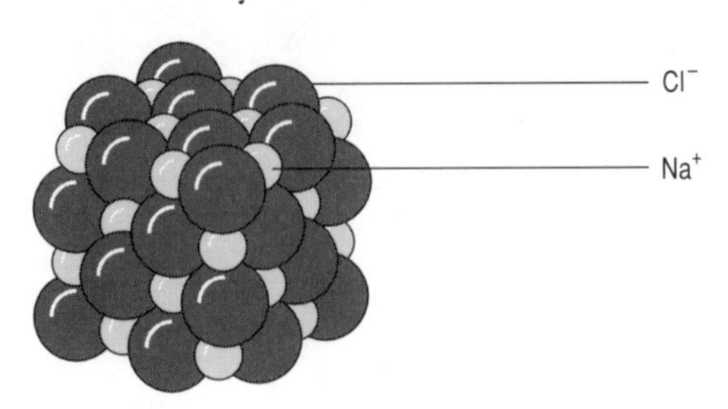

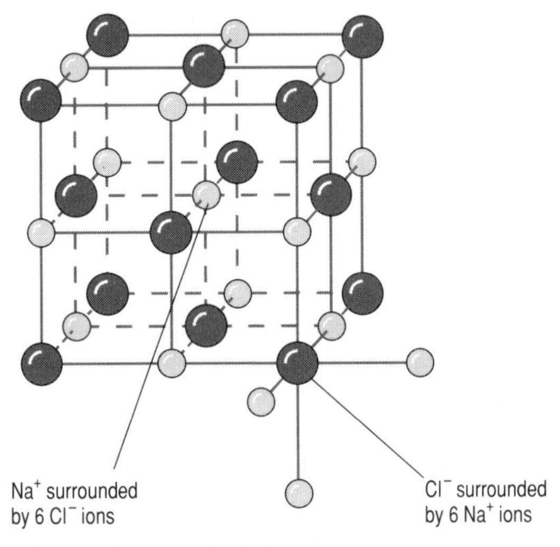

Figure 1.8.3 A sodium chloride lattice

Properties of ionic compounds

Ionic compounds have the following properties:

- They are usually solids at room temperature, with high melting points. This is due to the strong electrostatic forces of attraction holding the crystal lattice together. A lot of energy is therefore needed to separate the ions and melt the substance.
- They are usually hard substances.
- They usually cannot conduct electricity when solid, because the ions are not free to move.
- If they dissolve at all, they mainly dissolve in water. This is because water molecules are able to bond with both the positive and negative ions, which causes the lattice to break up and, hence, keeps the ions apart.
- They usually conduct electricity when in the molten state or in aqueous solution. The forces of attraction between the ions have been broken and the ions are free to move. This allows an electric current to be passed through, for example, molten sodium chloride.

Table 1.8.1

	Valency					
	1		2		3	
Metals	Lithium	(Li^+)	Magnesium	(Mg^{2+})	Aluminium	(Al^{3+})
	Sodium	(Na^+)	Calcium	(Ca^{2+})	Iron	(Fe^{3+})
	Potassium	(K^+)	Copper	(Cu^{2+})		
	Silver	(Ag^+)	Zinc	(Zn^{2+})		
	Copper	(Cu^+)	Iron	(Fe^{2+})		
			Lead	(Pb^{2+})		
			Barium	(Ba^{2+})		
Non-metals	Fluoride	(F^-)	Oxide	(O^{2-})		
	Chloride	(Cl^-)	Sulphide	(S^{2-})		
	Bromide	(Br^-)				
	Hydrogen	(H^+)				
Groups of atoms	Hydroxide	(OH^-)	Carbonate	(CO_3^{2-})	Phosphate	(PO_4^{3-})
	Nitrate	(NO_3^-)	Sulphate	(SO_4^{2-})		
	Ammonium	(NH_4^+)				
	Hydrogencarbonate	(HCO_3^-)				

Formulae of ionic substances

Ionic compounds contain positive and negative ions, whose charges balance. For example, sodium chloride contains one Na^+ ion to every Cl^- ion – giving rise to the formula NaCl.

This method can be used to write down formulae which show the ratio of the number of ions present in any ionic compound.

The formula of magnesium chloride is $MgCl_2$. This formula is arrived at because each Mg^{2+} ion combines with two Cl^- ions, and once again the charges balance. The size of the charge on an ion is a measure of its **valency** or combining power.

Na^+ has a valency of 1, but Mg^{2+} has a valency of 2. Na^+ can bond (combine) with only one Cl^- ion, whereas Mg^{2+} can bond with two Cl^- ions.

Some elements, such as copper and iron, possess two ions with different valencies. Copper can form both the Cu^+ ion and the Cu^{2+} ion and therefore it can form two different compounds with chlorine – CuCl, copper(I) chloride, and $CuCl_2$, copper(II) chloride. Iron forms the Fe^{2+} and Fe^{3+} ions.

Table 1.8.1 shows the valencies of a series of ions you will meet in your study of chemistry.

Table 1.8.1 also includes groups of atoms which have net charges. For example, the nitrate ion is a single unit composed of one nitrogen atom which has combined with three oxygen atoms, and has one single negative charge. The formula, therefore, of magnesium nitrate would be $Mg(NO_3)_2$. You will notice that the 'NO_3' has been placed in brackets, with a 2 outside the bracket. This indicates that there are two nitrate ions present for every magnesium ion. The ratio of the atoms present is therefore:

$$Mg(NO_3)_2$$
$$\downarrow$$
$$1Mg : 2N : 6O$$

Quick Questions

1 Draw diagrams to represent the bonding in each of the following ionic compounds:
  a) Calcium oxide (CaO);
  b) Magnesium chloride ($MgCl_2$);
  c) Lithium fluoride (LiF);
  d) Potassium chloride (KCl).

2 **a)** Using the information in Table 1.8.1 write the formulae of:
  i) Iron(II) chloride;
  ii) Calcium phosphate;
  iii) Iron(III) chloride;
  iv) Silver oxide.

  b) Using the formulae from your answers to **a)**, write down the ratio of atoms present for each of the compounds.

1.9 Covalent bonding

Another way in which atoms can gain the stability of the inert gas electron structure is by sharing the electrons in their outer energy levels. This occurs between **non-metal** atoms, and the bond formed is called a **covalent bond**. The simplest example of this type of bonding can be seen by considering the hydrogen molecule, H_2.

Each hydrogen atom in the molecule has one electron. In order to obtain a full outer energy level and gain the electron configuration of the inert gas, helium, each of the hydrogen atoms must have two electrons. To do this, the two hydrogen atoms allow their outer energy levels to overlap, see Figure 1.9.1.

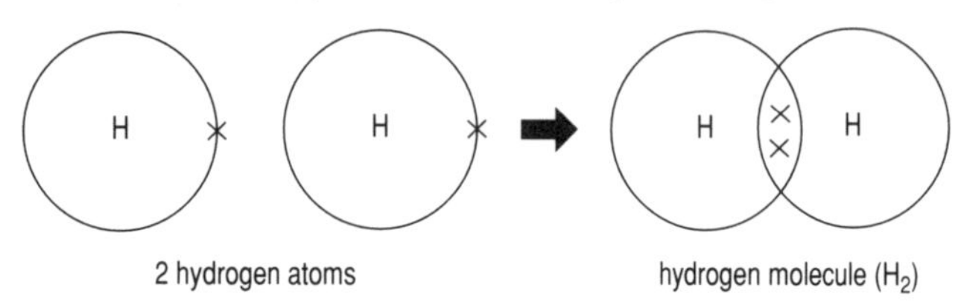

Figure 1.9.1 Electron sharing to form a hydrogen molecule

A molecule of hydrogen is formed from two hydrogen atoms, sharing a pair of electrons. This shared pair of electrons is attracted by the nucleus of each atom and is known as a single covalent bond. It can be represented by a single line.

Other covalent compounds

Methane (natural gas) is a gas whose molecules contain atoms of carbon and hydrogen. The electron structures are:

$$_6C\ 2,4 \qquad _1H\ 1$$

The carbon atom needs four more electrons to attain the electron configuration of the inert gas, neon. Each hydrogen atom only needs one more electron to form the electron configuration of helium. Figure 1.9.2 shows how the atoms gain these electron configurations by the sharing of electrons. Just as in the case of the examples given of ionic bonding, you will note that only the outer electron energy levels are shown.

Further examples of simple molecules are shown in the table below, as both structural formulae and dot/cross diagrams. Note that each line in the structural formulae represents one covalent bond. So each line represents one shared pair of electrons. Therefore, a double covalent bond involves the sharing of four electrons (2 pairs), as in an oxygen molecule, and a triple covalent bond involves the sharing of six electrons, as in nitrogen.

Table 1.9.1

Substance	Formula	Structural formula	Bonding scheme
chlorine	Cl_2	Cl—Cl	
water	H_2O	H—O—H	
ammonia	NH_3	H—N(—H)—H	
hydrogen chloride	HCl	H—Cl	
oxygen	O_2	O=O	
carbon dioxide	CO_2	O=C=O	
nitrogen	N_2	N≡N	

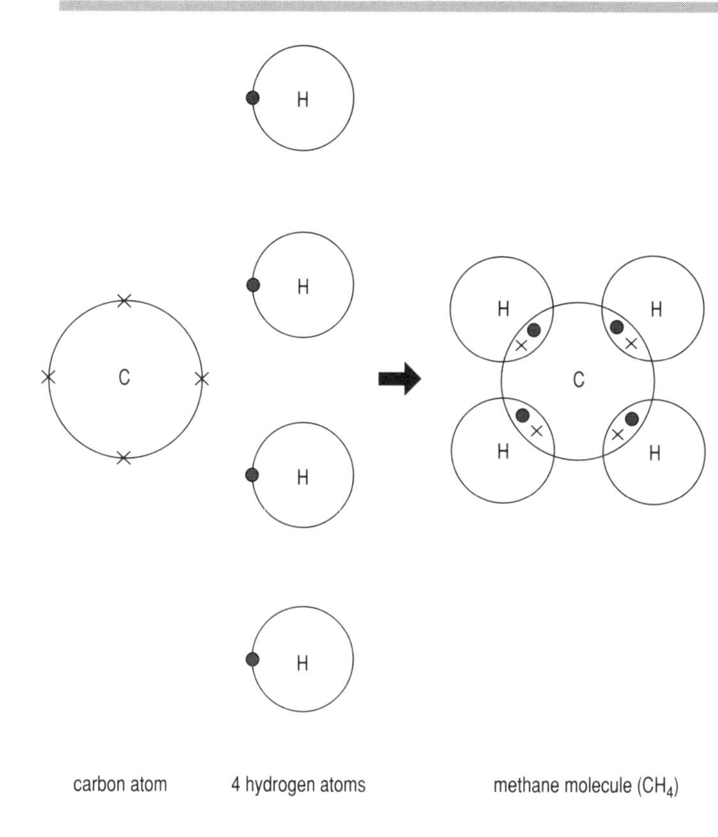

Figure 1.9.2 Bonding in a methane molecule

Covalent structures

Compounds containing covalent bonds have molecules whose structures can be classified as either **simple molecular** or **giant molecular**.

Simple molecular compounds are formed from only a few atoms. They have strong covalent bonds between the atoms within a molecule (**intramolecular bonds**), but have weak bonds between the molecules (**intermolecular bonds**). Examples of simple molecules are methane, iodine, water and ammonia. These weak bonds between molecules are known as van der Waals bonds, or forces, (see Figure 1.9.3) and these forces increase steadily with the increasing size of the molecule.

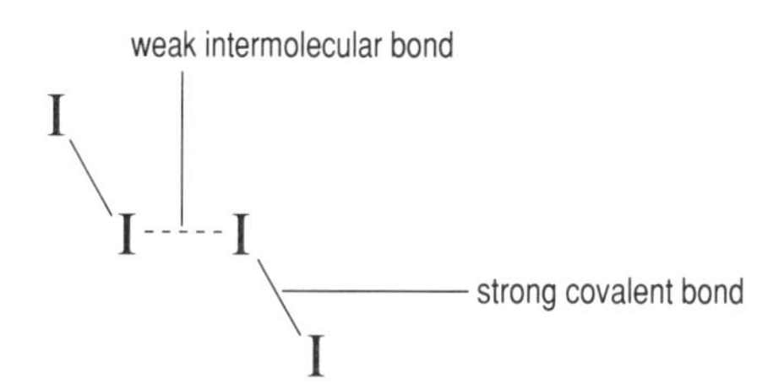

Figure 1.9.3 Strong covalent and weak intermolecular forces in iodine

Giant molecular structures contain many hundreds of thousands of atoms, joined by strong covalent bonds. Examples of substances showing this type of structure are diamond, graphite (see Section 1.10), silicon(IV) oxide (see Figure 1.9.4) and plastics (polymers) such as polyethene (see Section 1.10).

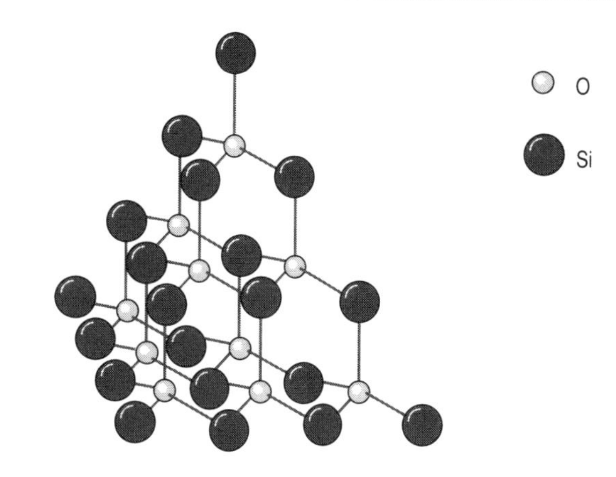

Figure 1.9.4 Silicon(IV) oxide structure

Properties of covalent compounds

Covalent compounds have the following properties:

- Simple molecular substances are usually gases, liquids or solids with low melting and boiling points. The melting points are low because of the weak intermolecular forces of attraction. Giant molecular substances have much higher melting points because the whole structure is held together by strong covalent bonds within the molecule.
- Generally they do not conduct electricity when molten or dissolved in water. This is because they do not contain ions. However, some molecules actually react with water to form ions. For example, hydrogen chloride produces aqueous hydrogen ions and chloride ions when it dissolves in water:

$$HCl(g) \xrightarrow{\text{water}} H^+(aq) + Cl^-(aq)$$

- Generally they do not dissolve in water unless they react with it to form ions. However, water is an excellent solvent and can interact with and dissolve some covalent molecules (e.g. sugars) better than others.

Quick Questions

1 With the aid of an example in each case, explain the following terms:
  a) Simple molecule;
  b) Single covalent bond;
  c) Double covalent bond.

2 Draw diagrams to represent the bonding in each of the following covalent compounds:
  a) Tetrafluoromethane (CF_4);
  b) Ethene (C_2H_4);
  c) Hydrogen sulphide (H_2S);
  d) Hydrogen fluoride (HF).

1.10 Giant structures

Giant structures have many hundreds of thousands of atoms, joined by strong covalent bonds. Each part is one enormous molecule sometimes called a **macromolecule**.

Allotropy

When an element can exist in more than one physical form in the same state it is said to exhibit allotropy (or polymorphism). Each of the different physical forms is called an **allotrope**. Allotropy is a relatively common feature of the elements in the Periodic Table. Graphite and diamond are called allotropes because they are both made up of the same element - carbon - and exist in the same physical states. Some examples of other elements which show allotropy are sulphur, tin and iron.

Allotropes of carbon

The difference in their structure helps to explain the difference in properties shown in Table 1.10.1.

Table 1.10.1 Physical properties of graphite and diamond

Property	Graphite	Diamond
Appearance	A dark grey, shiny solid	A colourless transparent crystal which sparkles in light
Electrical conductivity	Conducts electricity	Does not conduct electricity
Hardness	A soft material with a slippery feel	A very hard substance
Density (g cm^{-3})	2.25	3.51

Graphite

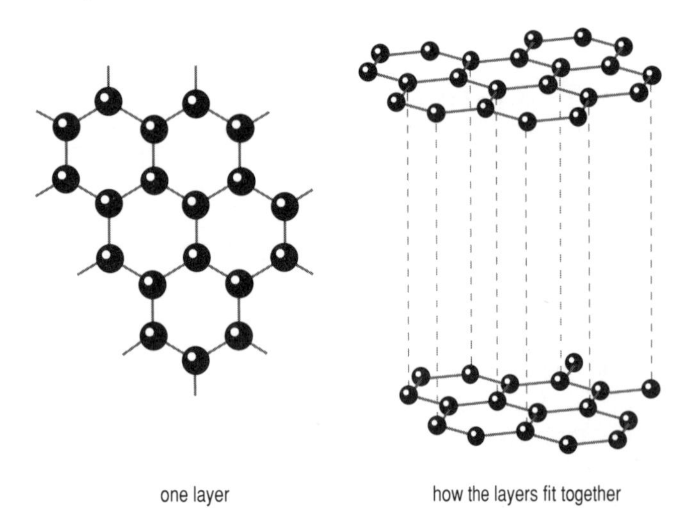

Figure 1.10.1 The structure of graphite

This is a layer structure in which, within each layer, each carbon atom is bonded to three others by strong covalent bonds. Each layer is therefore a giant molecule. Between these layers there are only weak forces of attraction (van der Waals forces) so the layers will easily pass over each other.

With only three covalent bonds formed between carbon atoms within the layers, an unbonded electron is present on each carbon atom. These 'spare' (or **delocalised**) electrons form electron clouds between the layers. It is because of these spare electrons that graphite conducts electricity.

Diamond

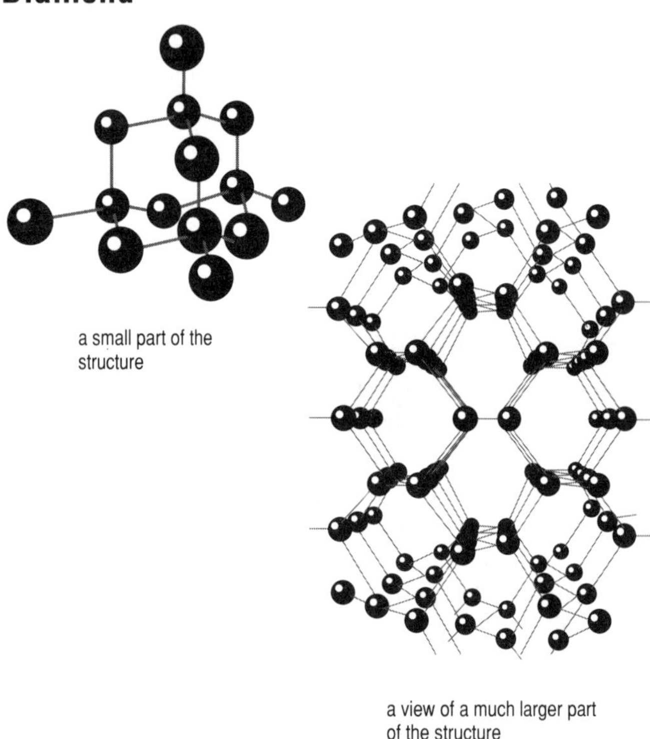

Figure 1.10.2 The structure of diamond

Each of the carbon atoms in the giant structure is covalently bonded to four others. They form a tetrahedral arrangement. This bonding scheme gives rise to a very rigid, three dimensional structure and accounts for the extreme hardness of the substance. All the outer energy level electrons of the carbon atom are used to form covalent bonds and so there are no electrons available to enable diamond to conduct electricity.

It is possible to manufacture both allotropes of carbon. Diamond is made by heating graphite to about 3000 °C at very high pressures. Diamond made by this method is known as industrial diamond. Graphite can be made by heating a mixture of coke and sand at a very high temperature in an electric arc furnace for about 24 hours.

Table 1.10.2 Uses of graphite and diamond

Graphite	Diamond
Pencils	Jewellery
Electrodes	Glass cutters
Lubricant	Diamond-studded saws
	Drill bits
	Polishers

Buckminsterfullerene

In 1985 an unusual new form of carbon was obtained. It was formed by the action of a laser beam on a sample of graphite.

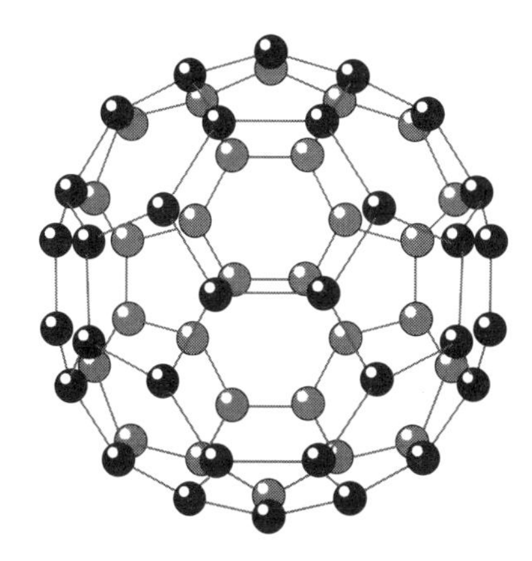

Figure 1.10.3 C_{60} buckminsterfullerene – a 'bucky ball'

This spherical structure is composed of 60 carbon atoms covalently bonded together. The discovery of further spherical forms of carbon, 'bucky balls', has lead to a whole new branch of inorganic carbon chemistry. It is thought that this type of molecule exists in chimney soot. Chemists have suggested that the large surface area of the bucky balls may mean that they have uses as catalysts.

Other forms of carbon

Charcoal and soot are also forms of carbon but, like fullerenes, they do not occur naturally. They are made by heating animal bones, wood or coal in a limited amount of air. Charcoal is used to decolorise solutions and as a deodoriser. In many parts of the world it is used as a fuel. Coke is another form of carbon which is manufactured to be used as a fuel. It is made by heating coal in the absence of air.

Plastics

Plastics such as polythene are a tangled mass of very long molecules in which the atoms are joined together by strong covalent bonds to form long chains called **polymers**. Molten plastics can be made into fibres by being forced through hundreds of tiny holes in a spinneret.

There are two different types of polymer, depending on what type of monomer is used to make them.

Thermoplastics, such as polythene, soften when heated but harden again when they are cooled. These plastics can be moulded into any shape. They contain long polymer chains which can slide around each other. Within the molecule the atoms are held by strong covalent forces, but the intermolecular forces between adjacent polymer chains are weak. Their shape, therefore, can be changed very easily.

Other plastics, such as those used to make pan handles and electrical plugs, must not soften when they are heated. These are **thermosetting plastics**. When thermosetting plastics are made, strong covalent bonds form between the chains - the chains become **cross-linked**.

The forces between adjacent chains are much stronger than in those plastics without cross links. This makes this type of polymer hard and rigid, even when hot. It also means that thermosetting plastics cannot be softened and so they cannot be remoulded.

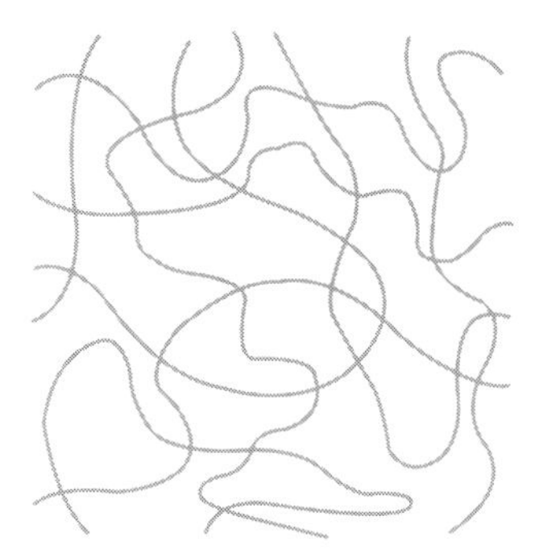

Figure 1.10.4 In thermosoftening plastic (or thermoplastic) there is no cross-linking

Figure 1.10.5 In thermosetting plastic the chains are cross-linked

Quick Questions

1. Make a summary table of the properties of the substances with covalent structures. Your table should include examples of both simple molecular and giant molecular substances.
2. With the aid of an example in each case, explain the following words and phrases:
   a) Intermolecular bond;
   b) Allotrope;
   c) Macromolecule;
   d) Thermoplastic;
   e) Thermosetting plastic.
3. It is said that both butter and margarine contain simple molecules. What properties do these two foods have which suggest that this statement is correct?

1.11 Metals, glasses and ceramics

Metals

Metals are amongst the most used, and most useful materials on Earth.

The different properties of metals are due to the type of chemical bond between the atoms within the metals - the **metallic bond**. This is another way in which atoms obtain a more stable electron structure. The electrons in the outer energy level of the atom of a metal move freely throughout the structure (they are delocalised) forming a mobile 'sea' of electrons. When the metal atoms lose these electrons they become positive ions. Therefore, metals consist of positive ions embedded in moving electrons (see Figure 1.11.1).

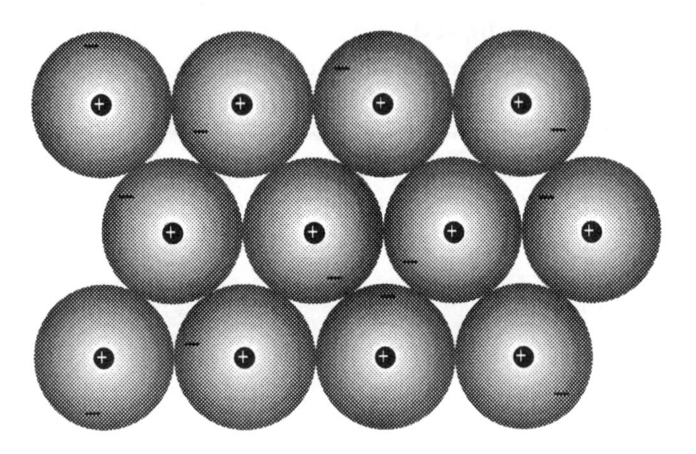

Figure 1.11.1 Metals consist of positive ions surrounded by a mobile sea of electrons

The negatively charged electrons attract all the positive metal ions and bond them together with strong electrostatic forces of attraction as a single unit. This is the metallic bond.

Properties of metals

Metals have the following properties:

- They usually have high melting and boiling points due to the strong attraction between the positive metal ions and the mobile sea of electrons.
- They conduct electricity due to the mobile sea of electrons within the metal structure. When a metal is connected in a circuit the mobile sea of electrons moves towards the positive terminal, whilst at the same time electrons are fed into the other end of the metal from the negative terminal.
- They are **malleable** and **ductile**. Unlike diamond, the bonds are not rigid but are still strong. If a force is applied to a metal, rows of ions can slide over one another. They reposition themselves and the strong bonds reform, as shown in Figure 1.11.2.

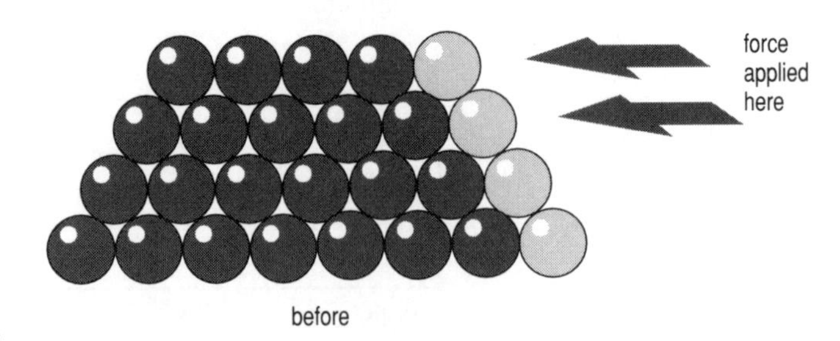

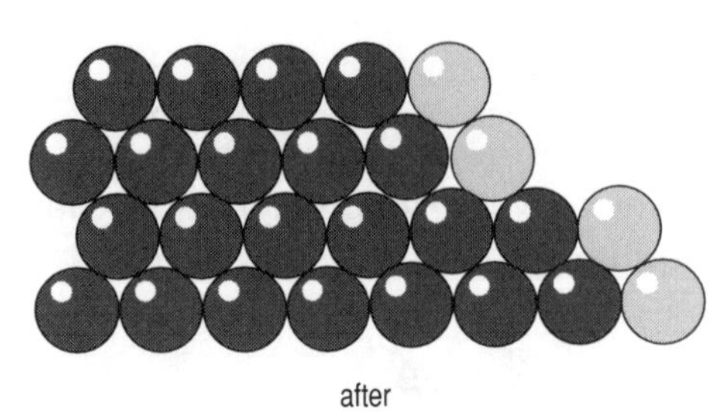

Figure 1.11.2 The position of the positive ions in a metal before and after a force has been applied

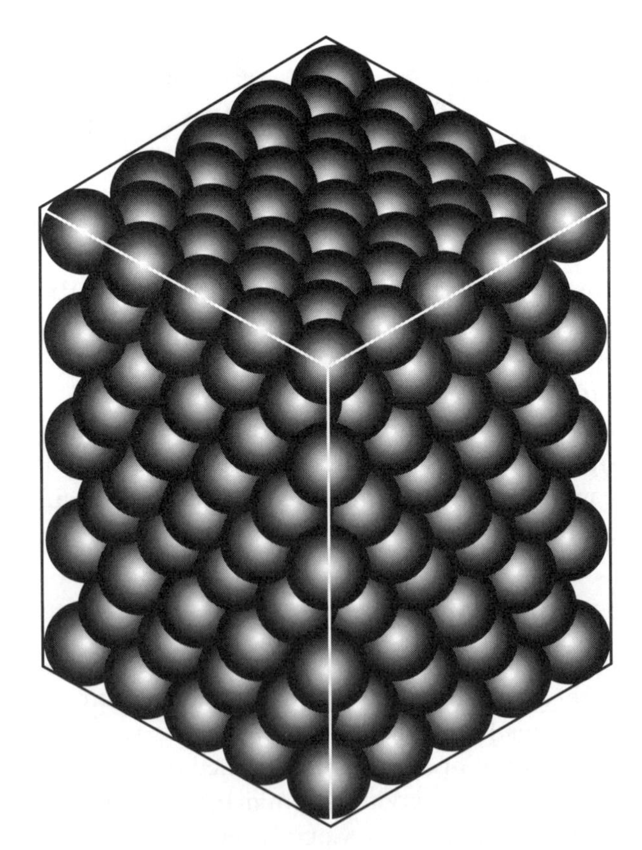

Figure 1.11.3 Arrangement of ions in the crystal lattice of a metal

- Malleable means that metals can be hammered into different shapes. Ductile means that the metals can be pulled out into thin wires.
- They have high densities. This arises because the atoms are very closely packed in a regular manner as can be seen in Figure 1.11.3.

Different metals show different types of packing and in doing so they produce the arrangement of ions.

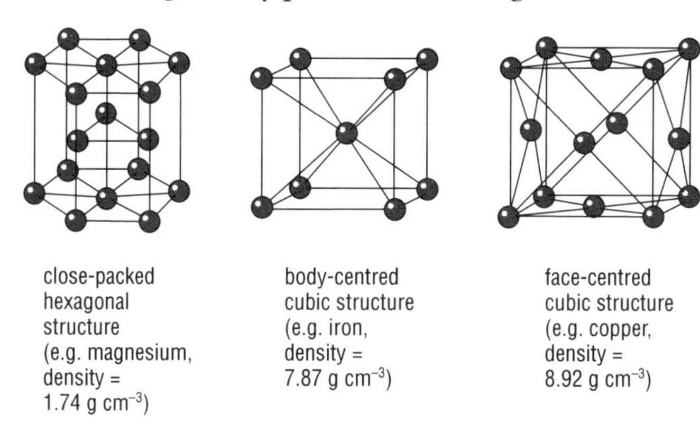

Figure 1.11.4 Relating structures to the density of metal

Glasses

Glasses are irregular giant molecular structures held together by strong covalent bonds. Glass can be made by heating silicon(IV) oxide with other substances until a thick viscous liquid is formed. As this liquid cools, the atoms present cannot move freely enough to return to their arrangement within the pure silicon(IV) oxide structure. Instead, they are forced to form a disordered arrangement as shown in Figure 1.11.5. Glass is called a supercooled liquid because it lacks the orderliness of a 'true' crystalline solid.

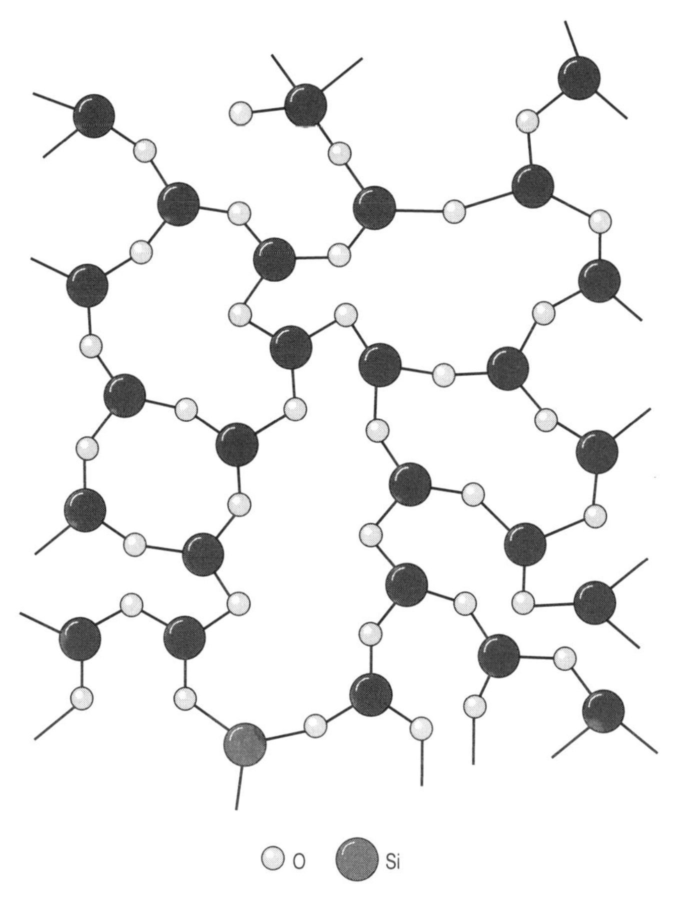

Figure 1.11.5 Two-dimensional structure of silicon(IV) oxide

The glass used in bottles and windows is called soda glass. This type of glass is made by heating a mixture of sand (silicon(IV) oxide), soda (sodium carbonate) and lime (calcium oxide). Pyrex is a borosilicate glass. It is made by incorporating some boron oxide into the silicon(IV) oxide structure so that silicon atoms are replaced by boron atoms. This type of glass is tougher than soda glass and more resistant to temperature changes. It is, therefore, used in the manufacture of cooking utensils and laboratory glassware.

Ceramic material

The word ceramic comes from the Greek meaning pottery, or 'burnt stuff'. Clay dug from the ground contains a mixture of several materials. The main one is a mineral called kaolinite, $Al_2Si_2O_5(OH)_4$, in which the atoms are arranged in layers in a giant structure. The clay can be moulded when it is wet because the kaolinite crystals move over one another. However, when it is dry, the clay becomes rigid because the crystals stick together.

During firing in a furnace the clay is heated to a temperature of 1000 °C. A complicated series of chemical changes takes place, new minerals are formed and some of the substances in the clay react to form a type of glass. The material produced at the end of the firing, the **ceramic**, consists of many minute mineral crystals bonded together with glass.

Quick Questions

1 a) Explain the terms:
   i) Malleable;
   ii) Ductile;
   iii) Glass;
   iv) Ceramic material.
 b) Explain why metals are able to conduct heat and electricity.
 c) Explain why metals generally have high densities.
 d) Explain why the melting point of magnesium (649 °C) is much higher than the melting point of sodium (97.9 °C).

2 Draw up a table to summarise the properties of the different types of substances you have met in this section. Your table should include examples from ionic, covalent (simple and giant), metals, ceramics and glasses.

3 The mineral kaolinite, $Al_2Si_2O_5(OH)_4$, is found in clay. Write down the ratio of the atoms present in this substance.

Section One: Examination Questions

1. (a) The diagram shows a burning candle.
Write beside each letter **A**, **B** and **C** to show if the candle wax is *solid*, *liquid* or *gas*.

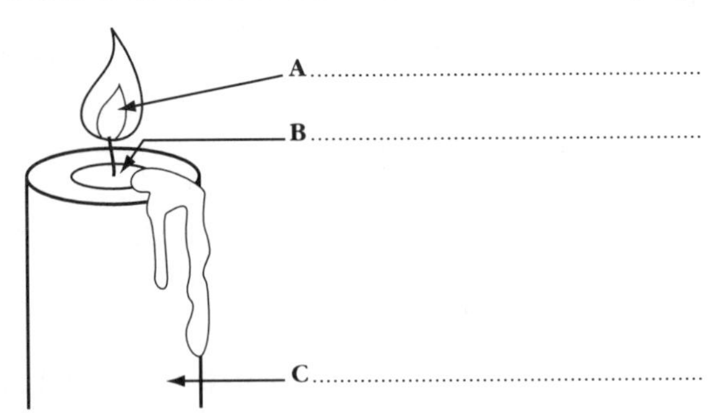

[3]

(b) (i) The first box below shows the particles in a solid metal. Complete the other boxes to show the particles when the metal is a liquid and when it is a gas.

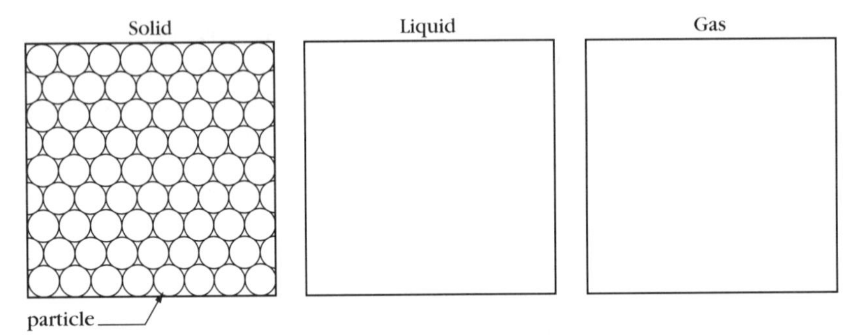

[3]

(ii) When the solid metal is heated, it expands. Explain what happens to the particles as the solid metal expands. [3]

(SEG Double Award, Foundation, 1998 Specimen)

2. Cars use both air and petrol in their engines.

(a) Put each of the gases in the air under the correct heading in the table. [2]

Element	**Compound**

(b) Petrol is a mixture of hydrocarbons.

(i) What is the meaning of the word *mixture*? [1]

(ii) Explain the meaning of the word *hydrocarbon*. [2]

(c) (i) Which gas in the air reacts with petrol in the engine? [1]

(ii) Name **two** substances formed when petrol burns in air. [2]

(SEG Double Award, Foundation, 1998 Specimen)

3. Golf club shafts need to be strong and reasonably stiff. They are made from wood, metal or composite materials such as carbon-fibre reinforced plastic.

(a) What do you understand by the term 'composite material'? [2]

Some properties of selected materials are given below.

Material	Density g/cm^3	Relative Strength	Relative Stiffness
Wood	0.6	1	20
Steel	7.8	10	105
Aluminium	2.7	2	35
Glass-fibre reinforced plastic	1.9	15	10
Carbon-fibre reinforced plastic	1.6	18	100
Nylon	1.1	0.8	1.5

(b) Which material from the list above would be **least** suitable for the shafts of golf clubs? Give a reason for your choice. [2]

(c) Cricket, hockey and hurling still use bats or sticks made from wood. Suggest **one** reason why other materials tend not to be used. [1]

(NICCEA Double Award, Higher, 1998 Specimen)

4. Rock salt contains insoluble solids and the soluble salt, sodium chloride. The following processes are needed to separate sodium chloride from rock salt.

addition of water   crystallisation   evaporation   filtration   stirring

Put each process in the correct order.
Explain the purpose of each process. [5]

(SEG Double Award, Higher, 1998 Specimen)

5. A student was asked to separate salt from a mixture of sand and salt. The separation was carried out in three stages.

Stage 1: the mixture is added to hot water and stirred.

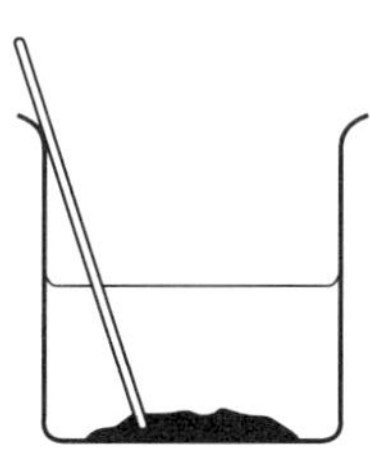

(a) What is the purpose of this stage? [1]

Stage 2: the mixture is now filtered and the residue washed with water.

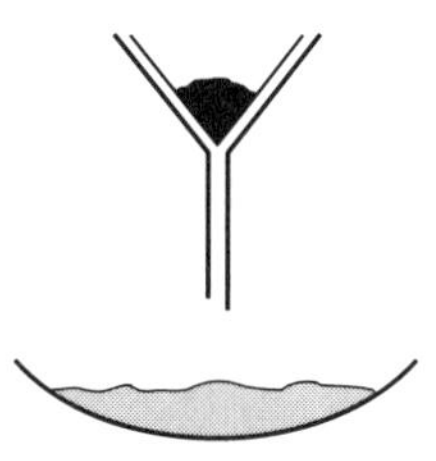

(b) Name the dissolved substance in the filtrate. [1]

(c) Why is the residue washed with water? [1]

(d) Describe how you would obtain a solid from the filtrate in **Stage 2**. [2]

(NICCEA Double Award, Higher, 1998 Specimen)

6. The diagram represents the arrangement of electrons in a magnesium atom.

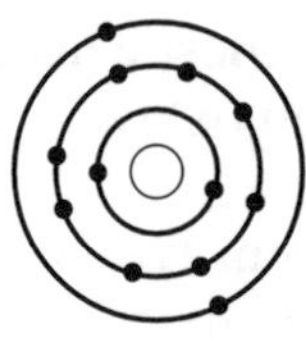

(a) Complete the table.

	number of			*electron arrangement*
	protons	*neutrons*	*electrons*	
magnesium-24				2,8,2
oxygen-16		8	8	

[3]

(b) Magnesium oxide contains ionic bonding.
Explain fully, in terms of transfer of electrons and the formation of ions, the changes which occur when magnesium oxide is formed from magnesium and oxygen atoms. [4]

(c) Sodium chloride and magnesium oxide have similar crystal structures and both contain ionic bonding. The melting points of sodium chloride and magnesium oxide are 800 °C and 2800 °C respectively. Suggest why the melting point of magnesium oxide is much higher than the melting point of sodium chloride.
(Sodium chloride contains Na^+ and Cl^- ions.) [3]

(MEG, Higher, 1998 Specimen)

7. This question is about sodium chloride (common salt) which is an important chemical. Sodium chloride can be made by burning sodium in chlorine gas.

(a) How could you prove by a **chemical test** that a test tube contains chlorine gas? [2]

(b) Name an element with similar chemical properties to chlorine. (You might find it helpful to refer to the periodic table on page 135.) [1]

(c) Write a word equation for the reaction between sodium and chlorine. [1]

(d) Complete the diagrams to show the electronic structure of a sodium atom and a chlorine atom.

sodium atom

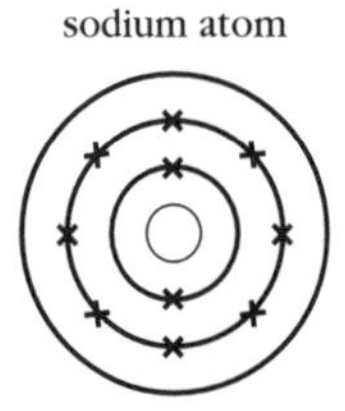

chlorine atom

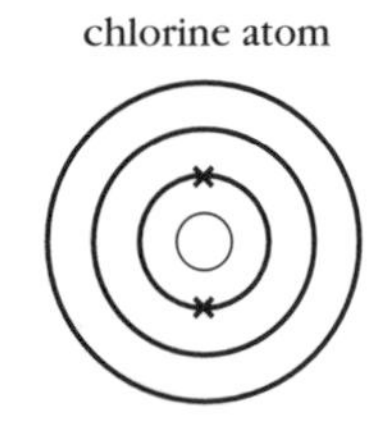

[2]

(e) How does a sodium atom change into a sodium ion? [2]

(f) The apparatus shown below can be used to electrolyse sodium chloride solution.

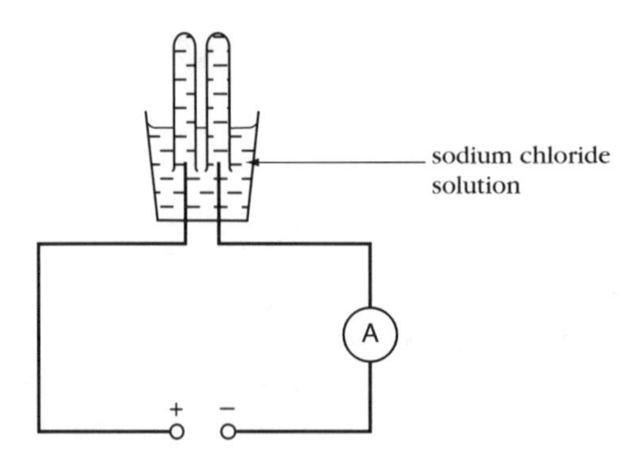

Products from the electrolysis of sodium chloride solution (brine) can be used to make other useful materials.
Name the product formed at
(i) the positive electrode
(ii) the negative electrode
Describe how these products are used to form useful materials. [7]
(NEAB, Foundation, 1998 Specimen)

8. A motorist called at a petrol filling station to get petrol, to check the oil level in the engine and to check the tyre pressures.

(a) (i) Explain fully, in terms of particles, why the motorist could smell the petrol as it was being put into the tank. [3]
(ii) Explain why the smell is more noticeable on a warm summer's day than during the winter. [2]
(b) After filling up with petrol, the motorist checked the oil level. Some oil was needed but as it was being added some was spilled on the ground. Explain, in terms of particles/kinetic theory, why it would take much longer for the oil spill to disappear than a small petrol spill. [3]
(c) The motorist then checked the tyre pressures. One of the tyres required some air. The car was then driven along a motorway to the next town. On checking the tyres again it was found that the pressure in each tyre had changed. Explain fully why the pressure had changed. [4]
(NICCEA, Higher, 1998 Specimen)

9. Use words from the list below to fill in the spaces.

gas liquid solid

(a) A .... takes the shape of a container into which it is poured. [1]
(b) When placed in a syringe a .... can be squeezed so that it takes up much less space. [1]
(c) A .... has a definite shape. [1]
(d) When released into a room a .... will spread out to fill the whole room. [1]
(NEAB, Foundation, 1998 Specimen)

10. Five different particles, labelled **A**, **B**, **C**, **D** and **E** are shown below:

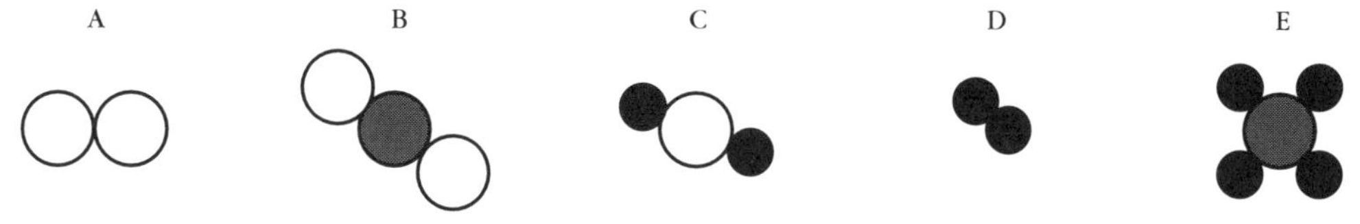

(i) How many atoms are there in one particle of compound E? [1]
(ii) List the particles in the diagram which are compounds. [3]
(iii) Write down the chemical formula for B. [1]
(WJEC, Foundation, 1998 Specimen)

11. There are millions of different substances that make up our world. All these substances are made from chemical elements.

(a) About how many different elements have been discovered? (The periodic table on page 135 may help you to answer this question.) [1]

(b) What is an element? [1]

(c) Many substances are compounds. What is a compound? [1]

(NEAB Coordinated, Foundation, 1995)

12. This question is about magnesium and its compounds.

(a) The bonding in magnesium is metallic.

(i) Draw a diagram to illustrate metallic bonding. [3]

(ii) Explain why magnesium is a good conductor of electricity. [1]

(iii) Use your understanding of metallic bonding to explain why metals can be pulled into wires. [1]

(b) (i) Draw a diagram to show the arrangement of the electrons in magnesium oxide, and show the charges on the ions. [3]

(ii) Suggest a reason why magnesium oxide is used to line the inside of furnaces. [1]

(NICCEA Double Award, Higher, 1998 Specimen)

13. **(a)** Lithium has an atomic number of 3 and a mass number of 7. This is often represented by the symbol

$$^{7}_{3}Li$$

(i) State the number of neutrons, protons and electrons which make up a neutral atom of this element.

(ii) Sometimes atoms of the same element occur with a different number of neutrons in the nucleus. What do we call such atoms?

(iii) Using the symbolism at the start of this question, how would you represent an atom of lithium which contained only 3 neutrons?

$$^{\dots\dots}_{\dots\dots}Li$$

(iv) Atoms of lithium form ions by the loss of one electron. Write the formula of a lithium ion. [7]

(b) (i) Give the electronic structure of a single, uncombined atom of chlorine.

(ii) Draw, showing only the outer electrons of each atom, the structure of a chlorine molecule, Cl_2. [2]

(c) Show what happens to the electronic structure of an atom of chlorine when it forms a chloride ion and say whether the atom has been oxidised, reduced or neither of these. [2]

(London Double Award, Higher, 1998 Specimen)

Section TWO

2.1 The periodic table

The **periodic table** was initially devised in 1869 by the Russian, Dimitri Mendeleev, who was Professor of Chemistry at St Petersburg University. His periodic table was based on the chemical and physical properties of the 60 elements that had been discovered at that time.

Mendeleev's periodic table has since been modified in the light of work carried out by Ernest Rutherford and Henry Mosley. Discoveries about sub-atomic particles led them to realise that the elements should be arranged by atomic number. In the modern periodic table, the 112 known elements are arranged in order of increasing atomic number.

The elements which have similar chemical properties are found in the same columns or **groups**. There are 8 groups of elements. The first column is called group 1; the second, group 2 and so on up to group 7. The final column in the periodic table is called group 0. Some of the groups have been given names:

Group 1: The alkali metals;
Group 2: The alkaline earth metals;
Group 7: The halogens;
Group 0: Inert gases or noble gases.

Between groups 2 and 3 is the block of elements known as the **transition elements**.

The horizontal rows are called **periods**, these are numbered 1 to 7 going down the left of the periodic table.

The periodic table can be divided into two as shown by the bold line in Figure 2.1.1. The elements to the left of this line are metals and those on the right are **non-metals**. The elements which lie on this dividing line are known as **metalloids**, for example, carbon (as graphite), silicon and germanium. These elements behave in some ways as metals and in others as non-metals.

Electron structures and the periodic table

The number of electrons in the outer energy level has been established. It can be seen that it corresponds with the number of the group in the periodic table in which the element is found. For example, the elements shown in Table 2.1.1 below have one electron in their outer energy level and they are all found in group 1.

Table 2.1.1

Element	Symbol	Atomic number	Electron structure
Lithium	Li	3	2.1
Sodium	Na	11	2.8.1
Potassium	K	19	2.8.8.1

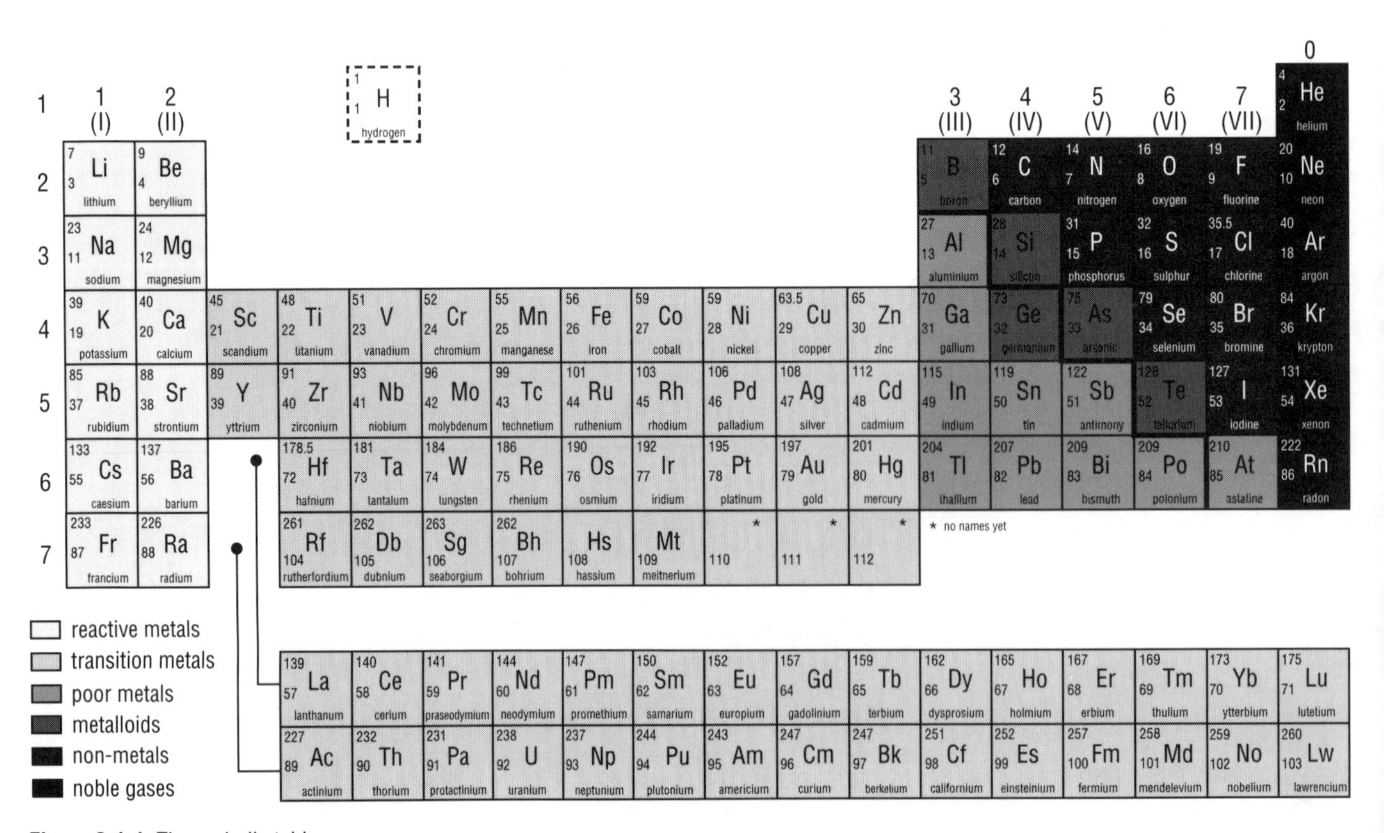

Figure 2.1.1 The periodic table

The elements in Table 2.1.2 below have two electrons in their outer energy level and they are found in group 2.

Table 2.1.2

Element	Symbol	Atomic number	Electron structure
Beryllium	Be	4	2.2
Magnesium	Mg	12	2.8.2
Calcium	Ca	20	2.8.8.2

The elements in group 0, however, are an exception to this rule as they have 2 or 8 electrons in their outer energy level. The outer electrons are mainly responsible for the chemical properties of any element, and therefore, elements in the same group have similar chemical properties.

Group 1 – the alkali metals

Group 1 consists of the six metals lithium, sodium, potassium, rubidium, caesium and the radioactive element francium. Lithium, sodium and potassium are commonly available in school. They are all very reactive metals and they are stored under oil to prevent them coming into contact with water or air.

These three metals have the following properties:

- They are good conductors of electricity and heat.
- They are soft metals.
- They are metals with low densities.
- They have shiny surfaces when freshly cut with a knife.
- They burn in oxygen or air, with characteristic flame colours, to form white, solid oxides. For example, lithium reacts with the oxygen in air to form white lithium oxide, according to the equation:

$$\text{lithium} + \text{oxygen} \longrightarrow \text{lithium oxide}$$
$$4Li(s) + O_2(g) \longrightarrow 2Li_2O(s)$$

These oxides all dissolve in water to form alkaline solutions of the metal hydroxide.

$$\text{lithium oxide} + \text{water} \longrightarrow \text{lithium hydroxide}$$
$$Li_2O(s) + H_2O(l) \longrightarrow 2LiOH(aq)$$

They react vigorously with water to give an alkaline solution of the metal oxides as well as hydrogen gas. For example,

$$\text{potassium} + \text{water} \longrightarrow \text{potassium hydroxide} + \text{hydrogen}$$
$$2K(s) + 2H_2O(l) \longrightarrow 2KOH(aq) + H_2(g)$$

Considering the group as a whole the further down you go the more reactive the metals become. So lithium is the least reactive and francium is, therefore, the most reactive of the group 1 metals.

Such gradual changes we call trends. Trends are useful to chemists as they allow predictions to be made about elements we have not observed in action.

Table 2.1.1 shows the electron structure of the first three elements of group 1. When these elements react they lose their single outer electron and in doing so become more stable because they obtain the electron structure of an inert gas.

When sodium loses its outer electron it requires energy to overcome the electrostatic attractive forces between the outer electron and the positive nucleus.

$$Na \longrightarrow Na^+ + e^-$$

Potassium is more reactive than sodium or lithium because less energy is required to remove the outer electron from its atom than for either lithium or sodium. This is because as you go down the group, the size of the atoms increases and the outer electron gets further away from the nucleus, hence becoming easier to remove.

Group 2 – the alkaline earth metals

Group 2 also consists of six metals. Magnesium and calcium are generally available in school. These metals have the following properties. They:

- Are harder than those in group 1.
- Are silvery-grey coloured when pure and clean, but tarnish quickly, however, when left in air due to the formation of a metal oxide.
- Are good conductors of heat and electricity.
- Burn in oxygen or air with characteristic flame colours to form solid white oxides.
- For example,

$$\text{magnesium} + \text{oxygen} \longrightarrow \text{magnesium oxide}$$
$$2Mg(s) + O_2(g) \longrightarrow 2MgO(s)$$

- React with water, but they do so much less vigorously than the elements in group 1.

  For example,

$$\text{calcium} + \text{water} \longrightarrow \text{calcium hydroxide} + \text{hydrogen}$$
$$Ca(s) + 2H_2O(l) \longrightarrow Ca(OH)_2(aq) + H_2(g)$$

Flame colours

If a clean nichrome wire is dipped into a metal compound and then held in the hot part of a Bunsen flame, the flame can become coloured.

Certain metal ions may be detected in their compounds by observing their characteristic flame colours. See Table 2.1.3 on the next page.

Table 2.1.3

Metal	Flame colour
Lithium	Crimson
Sodium	Golden yellow
Potassium	Lilac
Calcium	Brick red
Strontium	Crimson
Barium	Apple green
Lead	Blue/white
Copper	Green

Group 7 – the halogens

Group 7 consists of the five elements fluorine, chlorine, bromine, iodine and the radioactive astatine. Of these five elements chlorine, bromine and iodine are generally available in school.

Table 2.1.4

Element	Symbol	Atomic number	Electron structure
Fluorine	F	9	2,7
Chlorine	Cl	17	2,8,7
Bromine	Br	35	2,8,18,7
Iodine	I	53	2,8,18,18,7

These elements:

- Are coloured.

Halogen	**Colour**
Chlorine	Pale green
Bromine	Red/brown
Iodine	Purple/black

- Exist as **diatomic molecules**, for example, Cl_2, Br_2 and I_2.
- Show a gradual change from a gas (Cl_2) through liquid (Br_2) to solid (I_2).

Displacement reactions

If chlorine is bubbled into a solution of potassium bromide then the less reactive halogen, bromine, is displaced by the more reactive halogen, chlorine.

$$\text{potassium bromide} + \text{chlorine} \longrightarrow \text{potassium chloride} + \text{bromine}$$

$$2KBr(aq) + Cl_2(g) \longrightarrow 2KCl(aq) + Br_2(aq)$$

The observed order of reactivity of the halogens, confirmed by similar displacement reactions, is found to be:

$$\xrightarrow[\text{decreasing reactivity}]{\text{chlorine}\quad\text{bromine}\quad\text{iodine}}$$

You will notice that, unlike the elements of groups 1 and 2, the order of reactivity decreases going down the group. Table 2.1.4 shows the electron configuration for chlorine to bromine. In each case, the outer energy level contains 7 electrons. When these elements react they gain one electron per atom to gain the stable electron configuration of an inert gas.

$$Cl + e^- \longrightarrow Cl^-$$

Chlorine is more reactive than bromine because the incoming electron is being more strongly attracted into the outer energy level of the smaller atom. The attractive force on it will be greater than in the case of bromine since the outer energy level of chlorine is closer to the nucleus. As you go down the group this outermost, incoming electron is further from the nucleus. It will, therefore, be attracted less strongly and the reactivity of the elements in group 7 will decrease down the group.

Group 0 – the inert gases

This is a most unusual group of non-metals. They were all discovered after Mendeleev had published his periodic table.

They:

- Are colourless gases.
- Exist as **monatomic molecules**, He, Ne, Ar, Kr, Xe, Rn.
- Are very unreactive.

No compounds of helium, neon or argon have ever been found. However, more recently a number of compounds of xenon and krypton with fluorine and oxygen have been produced.

These gases are so chemically unreactive because they have electron configurations which are so stable that they are difficult to change.

Table 2.1.5

Element	Symbol	Atomic number	Electron structure
Helium	He	2	2
Neon	Ne	10	2.8
Argon	Ar	18	2.8.8

They are so stable that other elements attempt to attain these electron structures during chemical reactions. You have probably seen this in your study of the elements of groups 1, 2 and 7.

Although unreactive, they have many uses. Argon, for example, is the gas used to fill light bulbs to prevent the tungsten filament reacting with air, while neon is used extensively in advertising signs and in lasers. Helium is separated from natural gas by the liquefaction of the other gases. The other inert gases are obtained in large quantities by the fractional distillation of liquid air.

Transition elements

This block of metals includes many metals you will be familiar with, including, copper, iron, nickel, zinc and chromium.

Transition metals:

- Are harder and stronger than the metals in groups 1 and 2.
- Have much higher densities than the metals in groups 1 and 2.
- Are less reactive metals.
- Form a range of brightly coloured compounds.
- Are good conductors of heat and electricity.
- Show catalytic activity as elements and compounds. For example, iron is used in the industrial production of ammonia gas (Haber Process). (See Section 3.14.)
- Form more than one simple ion. For example, copper forms Cu^+ and Cu^{2+} and iron forms Fe^{2+} and Fe^{3+}.

The position of hydrogen

Hydrogen is often placed by itself in the periodic table. This is because the properties of hydrogen are unique.

However, profitable comparisons can be made with the other elements. It is often shown at the top of either group 1 or group 7, but it cannot fit easily into the trends shown by either group - as shown in Table 2.1.6.

Table 2.1.6 Comparison of hydrogen with lithium and fluorine

Lithium	Hydrogen	Fluorine
Solid	Gas	Gas
Forms a positive ion	Forms positive or negative ions	Forms a negative ion
1 electron in outer energy level	1 electron in outer energy level	1 electron short of a full outer energy level
Loses 1 electron to form a noble gas configuration	Needs 1 electron to form a noble gas configuration	Needs 1 electron to form a noble gas configuration

Quick Questions

1. Write word and balanced chemical equations for the reactions between:
   a) Sodium and oxygen;
   b) Magnesium and water.
2. Write word and balanced chemical equations for the reactions between:
   a) Bromine and potassium iodide solution;
   b) Bromine and potassium chloride solution.
   If no reaction will take place write 'no reaction' and explain why.
3. Which groups in the periodic table contain:
   a) Only metals;
   b) Only non-metals;
   c) Both metals and non-metals?
4. Account for the fact that calcium is more reactive than magnesium.

2.2 Metal reactivity

Metal reactions

By carrying out reactions in the laboratory with metals and air, water and dilute acid, it is possible to produce an order of **reactivity** for the metals.

With acid

If a metal reacts with dilute hydrochloric acid, then hydrogen and the metal chloride are produced. For example, when zinc reacts with dilute hydrochloric acid an **effervescence** is observed which is caused by bubbles of hydrogen gas being formed. The other product of this reaction is the salt, zinc chloride.

$$\text{zinc} + \text{hydrochloric acid} \longrightarrow \text{zinc chloride} + \text{hydrogen}$$
$$Zn(s) + 2HCl(aq) \longrightarrow ZnCl_2(aq) + H_2(g)$$

If similar reactions are carried out using other metals with acid, then an order of reactivity can be produced. This is known as a **reactivity series** and is shown in Table 2.2.1.

With air/oxygen

Many metals react directly with oxygen to form oxides. For example, calcium burns brightly in oxygen to form calcium oxide, a white powder.

$$\text{calcium} + \text{oxygen} \longrightarrow \text{calcium oxide}$$
$$2Ca(s) + O_2(g) \longrightarrow 2CaO(s)$$

However, a less reactive metal such as iron will react more slowly to give the oxide.

$$\text{iron} + \text{oxygen} \longrightarrow \text{iron(III) oxide}$$
$$2Fe(s) + 3O_2(g) \longrightarrow 2Fe_2O_3(s)$$

With water/steam

Reactive metals such as potassium, sodium and calcium react with cold water to produce the metal hydroxide and hydrogen gas. For example, the reaction of sodium with water produces sodium hydroxide and hydrogen.

$$\text{sodium} + \text{water} \longrightarrow \text{sodium hydroxide} + \text{hydrogen}$$
$$2Na(s) + 2H_2O(l) \longrightarrow 2NaOH(aq) + H_2(g)$$

Table 2.2.1 Order of reactivity of metals

Reactivity series	Reaction with dilute acid	Reaction with air/oxygen	Reaction with water	Ease of extraction
Potassium (K) Sodium (Na)	Produce H_2 with decreasing vigour	Burn very brightly and vigorously	Produce H_2 with decreasing vigour with cold water	Difficult to extract
Calcium (Ca)	↓	Burn to form an oxide with decreasing vigour		Easier to extract
Magnesium (Mg) (Aluminium (Al)*) Zinc (Zn) Iron (Fe)	↓	↓	React with steam with decreasing vigour	↓
Lead (Pb)	↓	React slowly to form the oxide	Do not react with cold water or steam	↓
Copper (Cu)	Do not react with dilute acids	↓	↓	Found as the element (native)
Silver (Ag) Gold (Au) Platinum (Pt)	↓	Do not react	↓	↓

Increasing Reactivity ↑

* Aluminium reacts very readily with the oxygen in the air, forming a layer of oxide on its surface. This often prevents any further reaction and disguises aluminium's true reactivity. This gives us the use of a light and strong metal.

The moderately reactive metals, magnesium, zinc and iron, react slowly with water. They will, however, react more rapidly with steam.

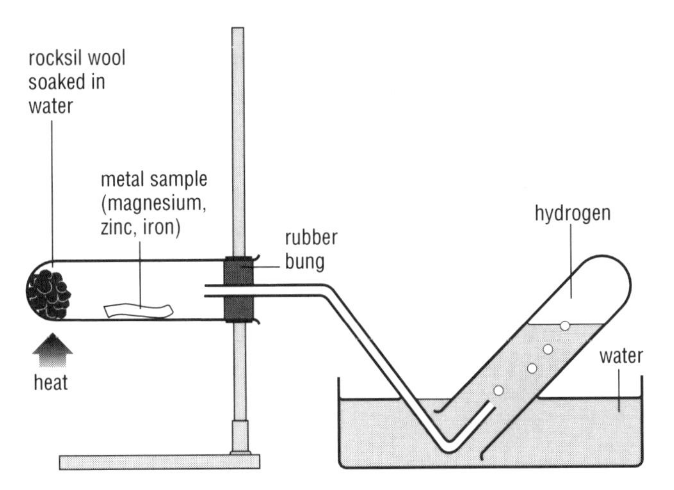

Figure 2.2.1 Apparatus used to investigate how metals such as magnesium react with steam

In their reaction with steam, the metal oxide and hydrogen are formed. For example, zinc produces zinc oxide and hydrogen gas.

$$\text{zinc} + \text{steam} \longrightarrow \text{zinc oxide} + \text{hydrogen}$$

$$Zn(s) + H_2O(g) \longrightarrow ZnO(s) + H_2(g)$$

Generally it is the unreactive metals that we find the most uses for. For example, the metals iron and copper can be found in many everyday objects such as copper pipes and wires and iron gates.

Both sodium and potassium are so reactive that they have to be stored under oil to prevent them coming into contact with water or air. Because they have low melting points and are good conductors of heat, they are used as coolants for nuclear reactors.

Quick Questions

1 a) Write word and balanced chemical equations for the reactions between:
i) Iron and dilute hydrochloric acid;
ii) Calcium and oxygen;
iii) Potassium and water.

b) For each use of a metal listed below, write a brief description of how its properties make it suitable for the job that it does.
i) Aluminium is used as window frames;
ii) Copper pans are used to make acidic foods such as jam and chutney;
iii) Food cans are made from steel which is covered with a layer of tin.

2 Bonium is a newly discovered metal. Bonium comes below aluminium, but above zinc, in the reactivity series. Predict how you would expect this newly discovered metal to react with:
a) Oxygen from the air;
b) Cold water;
c) Steam;
d) Dilute hydrochloric acid.
In each case name the products you would expect to be formed.

3 Use the following list of metals to answer the questions **a)–d)**:
iron, calcium, potassium, gold, aluminium, magnesium, sodium, zinc, platinum
a) Which of the metals will not react with oxygen to form an oxide?
b) Which of the metals will react violently with cold water?
c) Which of the metals has a protective coating on its surface?
d) Which of the metals reacts very slowly with cold water but extremely vigorously with steam?

2.3 Using the reactivity series

The reactivity series is useful in predicting how metals react.

Competition reactions in the solid state

If a more reactive metal is heated with the oxide of a less reactive metal, then it will remove the oxygen from it (as the oxide anion, O^{2-}). You can see from the reactivity series that iron is less reactive than aluminium. If iron(III) oxide is mixed with aluminium and the mixture is heated using a magnesium fuse, then a very violent reaction takes place as the aluminium takes the oxygen from the less reactive iron. It is a very exothermic reaction.

$$\text{iron(III) oxide} + \text{aluminium} \xrightarrow{\text{heat}} \text{aluminium oxide} + \text{iron}$$

$$Fe_2O_3(s) + 2Al(s) \longrightarrow Al_2O_3(s) + 2Fe(s)$$

After the reaction is over, a solid lump of iron is left along with a lot of white aluminium oxide powder.

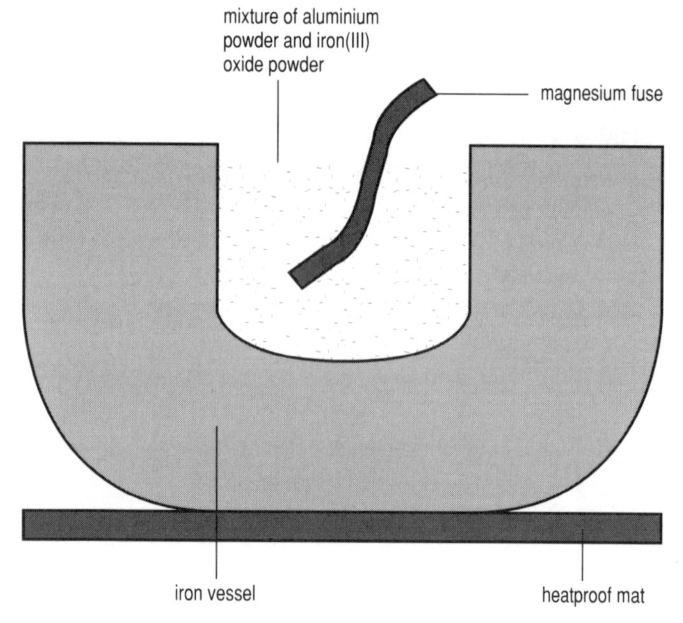

Figure 2.3.1

Figure 2.3.2 The oxygen is taken by the stronger (more reactive) metal

This is a **redox** reaction. A redox reaction is one in which both reduction and oxidation occur. The loss of oxygen is **reduction** whilst the gain of oxygen is **oxidation**. Therefore, the aluminium in this process has been oxidised. The iron, however, has lost its oxygen and has been reduced.

This particular reaction is known as the thermit reaction. Since large amounts of heat are given out during the reaction and the iron is formed in a molten state, this reaction is used to weld together damaged railway lines. Some metals, such as chromium and titanium, are extracted from their oxides using this type of competition reaction.

Competition reactions in aqueous solutions

In another reaction, metals compete with each other for anions. This type of reaction is known as a **displacement reaction**. As in the previous type of **competition reaction**, the reactivity series can be used to predict which of the metals will 'win'.

In a displacement reaction, a more reactive metal will displace a less reactive metal from a solution of its salt. Zinc is above copper in the reactivity series. When a piece of zinc metal is left to stand in a solution of copper(II) sulphate, the copper(II) sulphate slowly loses its blue colour as the zinc continues to displace the copper from the solution. The solution eventually becomes colourless zinc sulphate.

$$\text{zinc} + \text{copper(II) sulphate} \longrightarrow \text{zinc sulphate} + \text{copper}$$

$$Zn(s) + CuSO_4(aq) \longrightarrow ZnSO_4(aq) + Cu(s)$$

The ionic equation for this reaction is:

$$\text{zinc} + \text{copper ions} \longrightarrow \text{zinc ions} + \text{copper}$$

$$Zn(s) + Cu^{2+}(aq) \longrightarrow Zn^{2+}(aq) + Cu(s)$$

This is also a redox reaction, this time involving the transfer of electrons. Two electrons from the zinc metal transfer to the copper ions. The loss of electrons from the zinc metal is called oxidation. The gain of electrons by the copper ion is called reduction.

OIL	**RIG**
Oxidation **I**s **L**oss	**R**eduction **I**s **G**ain

of electrons

It is possible to confirm the reactivity series for metals using competition reactions of the types discussed in this section.

Identifying metal ions

When an alkali dissolves in water, it produces hydroxide ions ($OH^-(aq)$). It is known that most metal hydroxides are insoluble. So if hydroxide ions from a solution of an alkali are added to a solution of a metal salt, then an insoluble, often coloured, metal hydroxide is precipitated from solution.

For example:

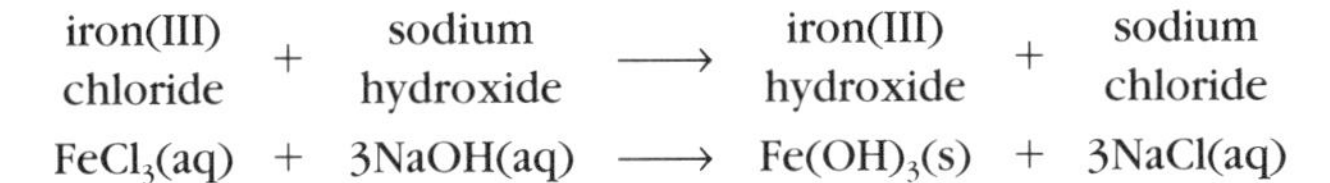

iron(III) chloride + sodium hydroxide ⟶ iron(III) hydroxide + sodium chloride

$$FeCl_3(aq) + 3NaOH(aq) \longrightarrow Fe(OH)_3(s) + 3NaCl(aq)$$

The ionic equation for this reaction is:

iron(III) ions + hydroxide ions ⟶ iron(III) hydroxide

$$Fe^{3+}(aq) + 3OH^-(aq) \longrightarrow Fe(OH)_3(s)$$

Table 2.3.1 Some of the colours of insoluble metal hydroxides

Name	Formula	Colour of hydroxide
Aluminium hydroxide	$Al(OH)_3$	White
Copper(II) hydroxide	$Cu(OH)_2$	Blue
Iron(II) hydroxide	$Fe(OH)_2$	Green
Calcium hydroxide	$Ca(OH)_2$	White
Iron(III) hydroxide	$Fe(OH)_3$	Brown (rust)
Zinc hydroxide	$Zn(OH)_2$	White

The colours of these insoluble metal hydroxides can be used to identify the metal cations which are present in solution.

Amphoteric hydroxides

The hydroxides of metals are basic and they react with acids to form salts (see Section 2.6). The hydroxides of some metals, however, will also react with strong bases, such as sodium hydroxide, to form soluble salts. Hydroxides of this type are said to be **amphoteric**.

For example:

zinc hydroxide + hydrochloric acid ⟶ zinc chloride + water

$$Zn(OH)_2(s) + 2HCl(aq) \longrightarrow ZnCl_2(aq) + 2H_2O(l)$$

and

zinc hydroxide + sodium hydroxide ⟶ sodium zincate

$$Zn(OH)_2(s) + 2NaOH(aq) \longrightarrow Na_2Zn(OH)_4(aq)$$

Other amphoteric hydroxides include lead hydroxide ($Pb(OH)_2$) and aluminium hydroxide ($Al(OH)_3$). This sort of behaviour can be used to help in the identification of these metal cations, as their hydroxides are soluble in strong bases.

It should be noted that the oxides of the metals used as examples above are also amphoteric.

Quick Questions

1. What do you understand by the following:
   a) Thermit reaction;
   b) Competition reaction;
   c) Redox reaction;
   d) Oxidation;
   e) Reduction;
   f) Amphoteric?
2. Predict whether or not the following will react:
   a) Magnesium + copper(II) oxide;
   b) Iron + aluminium oxide;
   c) Calcium + magnesium oxide.
   Complete the word equations, and write balanced chemical and ionic equations for those reactions which do take place.
3. Predict whether or not the following will react:
   a) Magnesium + calcium nitrate solution
   b) Iron + copper(II) nitrate solution
   c) Copper + silver nitrate solution
   Complete the word equations, and write balanced chemical and ionic equations for those reactions which do take place.
4. Write ionic equations for the reactions which take place to produce the metal hydroxides shown in Table 2.3.1.
5. Describe what you would see when sodium hydroxide is added slowly to a solution containing lead(II) nitrate.
6. Describe how you would distinguish between solutions containing:
   a) Aluminium and calcium ions;
   b) Iron(II) and iron(III) ions.
7. Magnesium can be reacted with steam using the apparatus shown in Figure 2.2.1. When gas **A** is collected, mixed with air and ignited it gives a small pop. A white solid **B** remains in the test tube when the reaction has stopped.
   a) Name and give the formula of gas **A**.
   b) i) Name the product formed when gas **A** burns in air.
   ii) Write a balanced chemical equation for this reaction.
   c) i) Name white solid **B**.
   ii) Write a balanced chemical equation to represent the reaction between magnesium and steam.
   d) Name two other metals which could be safely used to replace magnesium and produce another example of gas **A**.
   e) When magnesium reacts with dilute hydrochloric acid, gas **A** is produced again. Write a balanced chemical equation to represent this reaction and name the other product of this reaction.

2.4 Useful products from ores – 1

Extraction of metals from their ores

The majority of metals are too reactive to exist on their own in the Earth's **crust** and instead, they occur naturally in rocks as compounds in **ores**. These ores are usually carbonates, oxides or sulphides of the metal, mixed with impurities.

Some metals, such as gold and silver, occur 'native' as the free metal. They are very unreactive metals and have withstood the action of water and the atmosphere for many thousands of years without reacting and becoming compounds.

Large lumps of the ore are first crushed and ground up by very heavy machinery. Some ores are already fairly concentrated when mined. For example, haematite can contain over 80% of iron(III) oxide, Fe_2O_3. However, copper pyrites, for example, may only have 1% or less of the copper compound and these ores must be concentrated before the metal can be extracted. The method then used to extract the metal from its ore depends on the position of the metal in the reactivity series.

Electrolysis

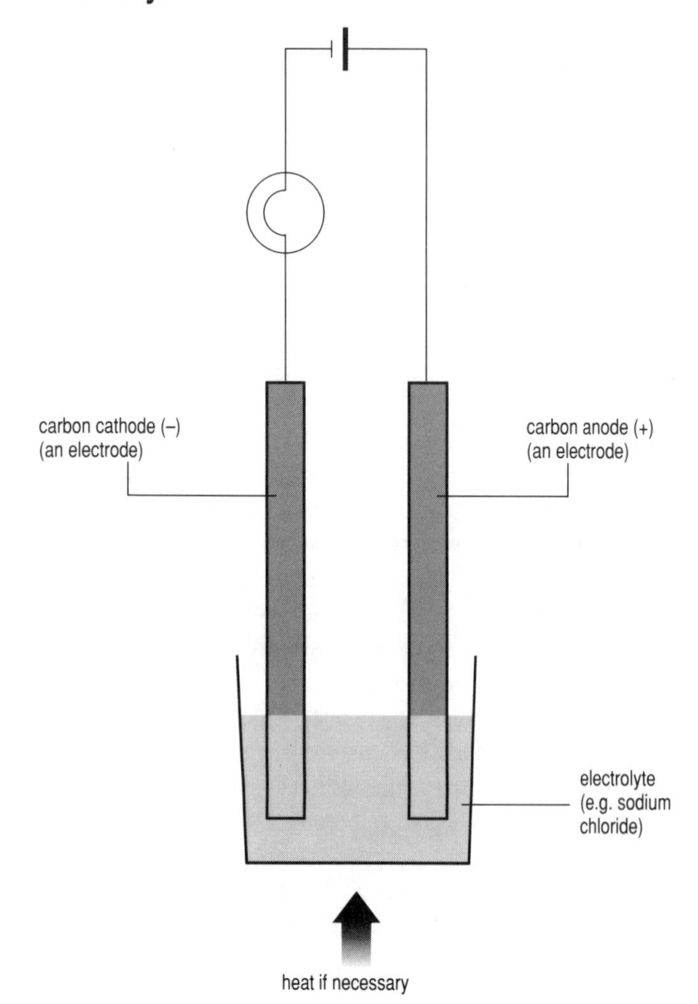

Figure 2.4.1 The important terms used in electrolysis

Electrolysis is the process of splitting up (decomposing) substances by passing an electric current through them. The substance that is decomposed is called the **electrolyte**, which conducts electricity when in the molten state or in solution. The electricity is carried through the electrolyte by ions. The electric current enters and leaves the electrolyte through **electrodes** (Figure 2.4.1) which are usually made of unreactive metals such as platinum or from the non-metal carbon.

The names given to the two electrodes are **cathode** - the negative electrode, and **anode** - the positive electrode. The process of electrolysis is a very important one in industry.

Extraction of aluminium

Aluminium is a reactive metal. It is obtained from its **ore**, bauxite, which contains aluminium oxide (Al_2O_3). Its high reactivity means it holds on to the oxygen it has combined with very strongly. The process developed to extract aluminium is called the Hall-Héroult process.

- Bauxite is first treated with sodium hydroxide to obtain pure aluminium oxide, by removing impurities such as iron(III) oxide and sand.
- The purified aluminium oxide is then dissolved in molten cryolite (Na_3AlF_6). Cryolite, a mineral found naturally in Greenland, is used to reduce the working temperature of the Hall-Héroult cell to between 800–1000 °C. The melting point of pure aluminium oxide is 2017 °C. Therefore, using the cryolite provides a considerable saving in the energy requirements.
- The molten mixture is then electrolysed in a cell similar to that shown in Figure 2.4.2.

The anodes of this process are blocks of graphite which are lowered into the molten mixture from above. The cathode is the graphite lining of the steel vessel which contains the cell.

Aluminium oxide is an ionic compound. When it is melted, the ions become mobile as the strong electrostatic forces of attraction between them are broken by the input of heat energy. During the electrolysis process the negatively charged oxide ions are attracted to the anode (the positive electrode). Here, they lose electrons and form oxygen gas (oxidation).

$$\text{oxide ions} \longrightarrow \text{oxygen molecules} + \text{electrons}$$
$$2O^{2-}(l) \longrightarrow O_2(g) + 4e^-$$

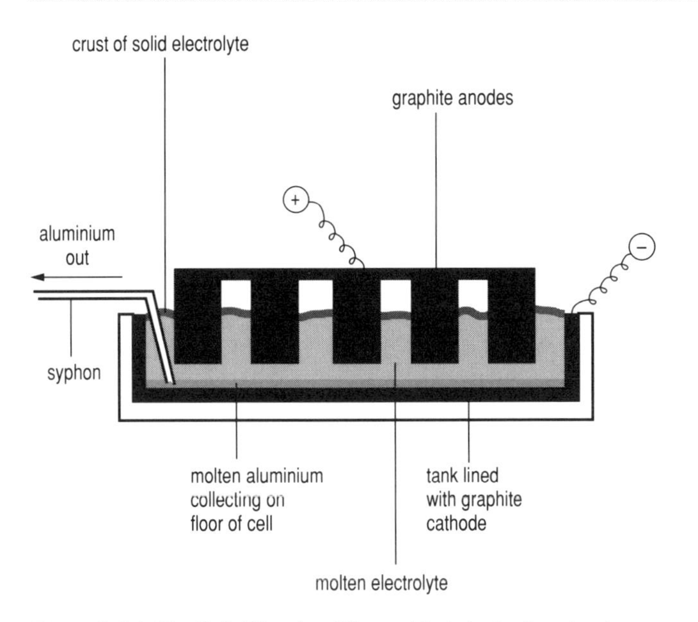

Figure 2.4.2 The Hall-Héroult cell is used in industry to extract aluminium by electrolysis

The positive aluminium ions are attracted to the cathode (the negative electrode). They gain electrons to form molten aluminium metal (reduction).

aluminium ions	+	electrons	⟶	aluminium metal
$Al^{3+}(l)$	+	$3e^-$	⟶	$Al(l)$

The overall reaction which takes place in the cell is:

$$\text{aluminium oxide} \xrightarrow{\text{electrolysis}} \text{aluminium} + \text{oxygen}$$
$$2Al_2O_3(l) \longrightarrow 2Al(l) + 3O_2(g)$$

The molten aluminium collects at the bottom of the cell and it is siphoned out at regular intervals.

Problems do arise, however, with the graphite anodes. At the working temperature of the cell, the oxygen liberated reacts with the graphite anodes to produce carbon dioxide.

carbon	+	oxygen	⟶	carbon dioxide
$C(s)$	+	$O_2(g)$	⟶	$CO_2(g)$

The anodes are burned away and must be replaced regularly. This is a continuous process in which vast amounts of electricity are used. Hydroelectric power, the cheapest form of electricity, is usually used for this process.

Uses of aluminum

Aluminium has many uses including bikes, cooking foil and overhead power cables, as well as in alloys such as duralumin which is used in the manufacture of aeroplane bodies.

Environmental issues

Environmental problems associated with the location of aluminium plants are concerned with:

- The effects of the extracted impurities which form a red mud.
- The fine cryolite dust which is emitted through very tall chimneys so as to minimise the effect on the surrounding area.
- The claimed link between environmental aluminium and Alzheimer's disease.

Quick Questions

1. Produce a flow chart to summarise the processes involved in the extraction of aluminium metal.
2. Explain the following terms in relation to the extraction of aluminium: anode, cathode, electrolysis, electrolyte, oxidation, reduction.
3. List five uses of aluminium metal, stating which property (or properties) is (are) important for each use.

2.5 Useful products from ores – 2

Metals towards the middle of the reactivity series, such as iron and zinc, may be extracted by reducing the metal oxide with the non-metal, carbon.

Iron

Iron is extracted mainly from its oxides, haematite (Fe_2O_3) and magnetite (Fe_3O_4), in a blast furnace.

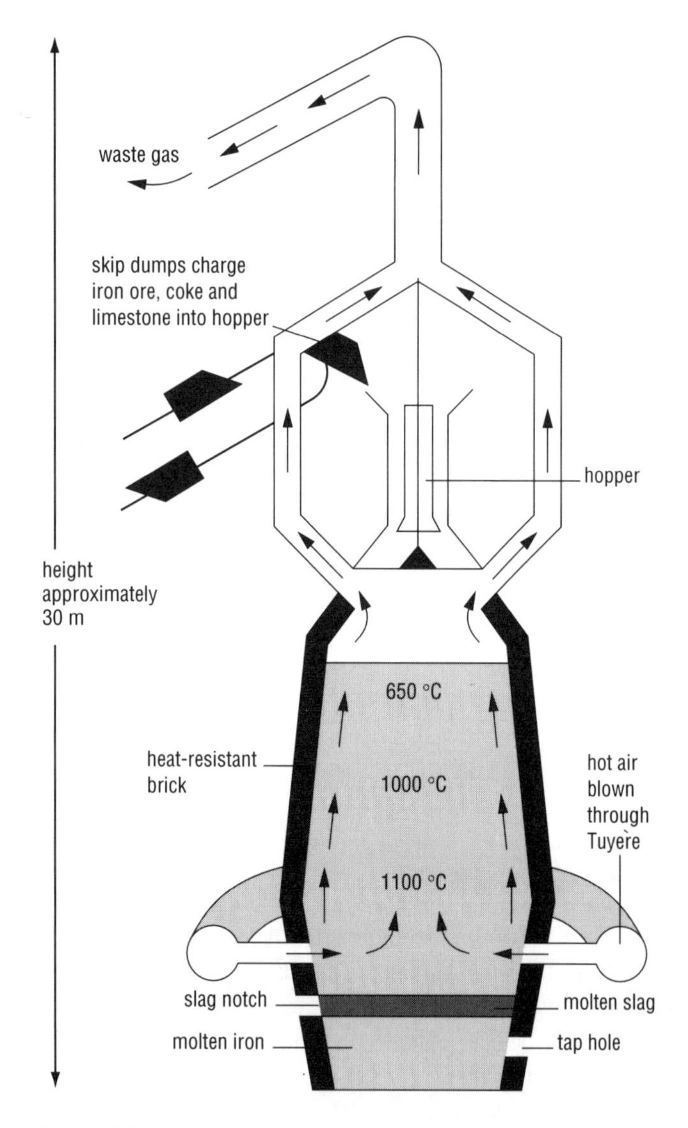

Figure 2.5.1 Cross section of a blast furnace

The blast furnace is a steel tower approximately 30 metres high, lined with heat-resistant bricks. It is loaded with the 'charge' of iron ore (usually haematite), coke (made by heating coal), and **limestone** (calcium carbonate). A blast of hot air is sent in near the bottom of the furnace through holes (tuyères). This makes the charge in the furnace glow as the coke burns in the preheated air.

$$\text{carbon} + \text{oxygen} \longrightarrow \text{carbon dioxide}$$
$$C(s) + O_2(g) \longrightarrow CO_2(g)$$

A number of chemical reactions then follow:

- The limestone begins to decompose.

$$\text{calcium carbonate} \longrightarrow \text{calcium oxide} + \text{carbon dioxide}$$
$$CaCO_3(s) \longrightarrow CaO(s) + CO_2(g)$$

- The carbon dioxide gas so produced then reacts with more hot coke higher up in the furnace, producing carbon monoxide.

$$\text{carbon dioxide} + \text{coke} \longrightarrow \text{carbon monoxide}$$
$$CO_2(g) + C(s) \longrightarrow 2CO(g)$$

- The carbon monoxide is a **reducing agent**. It rises up the furnace and reduces the iron(III) oxide ore. This takes place at a temperature of approximately 700 °C.

$$\text{iron(III) oxide} + \text{carbon monoxide} \longrightarrow \text{iron} + \text{carbon dioxide}$$
$$Fe_2O_3(s) + 3CO(g) \longrightarrow 2Fe(l) + 3CO_2(g)$$

The iron(III) oxide is behaving as an **oxidising agent**. The molten iron so produced trickles to the bottom of the furnace.

- The calcium oxide (an insoluble base) formed from the limestone reacts with acidic impurities including silicon(IV) oxide (sand), SiO_2, in the iron ore to form a liquid slag which is mainly calcium silicate.

$$\text{calcium oxide} + \text{silicon(IV) oxide} \longrightarrow \text{calcium silicate}$$
$$CaO(s) + SiO_2(s) \longrightarrow CaSiO_3(l)$$

This material also trickles to the bottom of the furnace, but because it is less dense than the molten iron, it floats on the top of it. The molten iron, as well as the molten slag, may be tapped off (run off) at intervals.

The waste gases, mainly nitrogen and oxides of carbon, escape from the top of the furnace and are used in a heat exchange process to heat in-coming air, helping to reduce the energy costs of the process. The slag is the other waste material. It is used by builders and road makers for foundations.

The extraction of iron is a **continuous process** and is much cheaper to run than an electrolytic method.

The iron which is obtained by this process is known as 'pig' or cast iron and contains about 5% carbon (as well as some other impurities). The name pig iron arises from the fact that if it is not subsequently converted into steel, the molten iron is poured into moulds called pigs. The iron produced by this process is brittle and hard, so has limited use. Gas cylinders are sometimes made of cast iron since they are unlikely to get deformed during their use.

The majority of the iron produced in the blast furnace is converted into different steel alloys such as manganese and tungsten steels, as well as stainless steel.

Extraction of titanium

Titanium metal cannot be extracted successfully by electrolysis. It is extracted from titanium(IV) chloride, $TiCl_4$, which is covalently bonded and so cannot conduct electricity. Instead it is extracted by reaction with a more reactive metal.

Titanium is a hard, silvery metal of low density. It is corrosion resistant and has many uses in the manufacture of strong, light alloys for use in aircraft, missile manufacture and car engines.

Titanium is found in the ore, rutile (TiO_2). The metal is extracted from these ores by the following processes:

- The ore is processed to obtain pure titanium(IV) oxide, TiO_2.
- The titanium(IV) oxide is converted to titanium(IV) chloride, $TiCl_4$, by reaction with hydrochloric acid.
- Titanium metal is obtained by a displacement reaction resulting from the reaction between the titanium(IV) chloride and the more reactive sodium or magnesium metals.

$$\text{titanium chloride} + \text{sodium} \longrightarrow \text{sodium chloride} + \text{titanium}$$

$$TiCl_4(l) + 4Na(l) \longrightarrow 4NaCl(s) + Ti(s)$$

Extraction of unreactive metals

Silver and gold are very unreactive metals. Silver exists mainly as silver sulphide, Ag_2S, (silver glance). The extraction involves treating the pulverised ore with sodium cyanide. Zinc is then added, to displace the silver from solution. The pure metal is obtained by electrolysis. It also exists, to a minor extent, 'native' (as a free element) in the Earth's crust. Gold is nearly always found native. It is also obtained in significant amounts during the electrolytic refining of copper, as well as during the extraction of lead.

Silver and gold, because of their resistance to corrosion, are used to make jewellery. Both these metals are used in the electronics industry because of their high electrical conductivity.

Quick Questions

1. How does the method used for extracting a metal from its ore depend on its position in the reactivity series?
2. 'It is true to say that almost all the reactions in which a metal is extracted from its ore are reduction reactions.' Discuss this statement with respect to the extraction of iron, aluminium and titanium.
3. Suggest a method which could be used to extract magnesium from magnesium chloride.

2.6 Acids and bases

Acids and alkalis

The word **acid** means sour and all acids possess this property. They are also:

- Soluble in water.
- Corrosive.

Alkalis are the chemical opposite of acids. They:

- Will remove the sharp taste from an acid.
- Have a soapy feel.

It would be too dangerous to taste a liquid to find out if it were acidic. Chemists use substances called **indicators** which change colour when they are added to acids or alkalis. Many indicators are dyes which have been extracted from natural sources. For example, litmus is a purple dye which has been extracted from lichens.

Table 2.6.1

Indicator	Colour in acid	Colour in alkali
phenolphthalein	colourless	pink
methyl orange	red	yellow
red litmus	red	blue
blue litmus	red	blue
methyl red	red	yellow

To find out how acidic or how alkaline a substance is we use another indicator, known as **universal indicator**, which is a mixture of many indicators. The colour shown by this indicator can be matched against a **pH scale** which runs from below 0 to 14 (see table on page 134). A substance with a pH of less than 7 is an acid. One with a pH of greater than 7 is alkaline and one with a pH of 7 is **neutral**. Water is the most common example of a neutral substance.

pH can also be measured using a pH meter. The electrode is placed into the solution and a pH reading is given on the digital display.

Strong and weak acids

Different acids have different pH values. A typical **strong acid** is hydrochloric acid. It is formed by dissolving hydrogen chloride gas in water. In hydrochloric acid, the ions formed separate completely.

$$\text{hydrogen chloride gas} \xrightarrow{\text{water}} \text{hydrogen ions} + \text{chloride ions}$$

$$HCl(g) \longrightarrow H^+(aq) \quad Cl^-(aq)$$

Any acid which behaves in this way is termed a strong acid. Both sulphuric and nitric acids are also strong acids.

All these acids have a high concentration of hydrogen ions in solution ($H^+(aq)$). Strong acids have a low pH. A **weak acid**, such as ethanoic acid which is found in vinegar, produces only a few hydrogen ions when it dissolves in water.

$$\text{ethanoic acid} \overset{\text{water}}{\rightleftharpoons} \text{hydrogen ions} + \text{ethanoate ions}$$

$$CH_3COOH(l) \rightleftharpoons H^+(aq) + CH_3COO^-(aq)$$

The $\rightleftharpoons$ sign means that the reaction is reversible. This means that if the ethanoic acid molecule breaks down to give hydrogen ions and ethanoate ions, then they will react together to reform the ethanoic acid molecule. This means that only a few hydrogen ions are present in the solution at a given time. Other examples of weak acids are citric acid found in oranges and lemons, carbonic acid found in soft drinks, sulphurous acid (acid rain), and ascorbic acid (vitamin C).

All acids when in aqueous solution produce hydrogen ions ($H^+(aq)$). To say that an acid is a strong acid does not mean it is concentrated. The strength of an acid tells you how easily it ionises to produce hydrogen ions. The concentration of an acid indicates the proportion of water and acid present in aqueous solution.

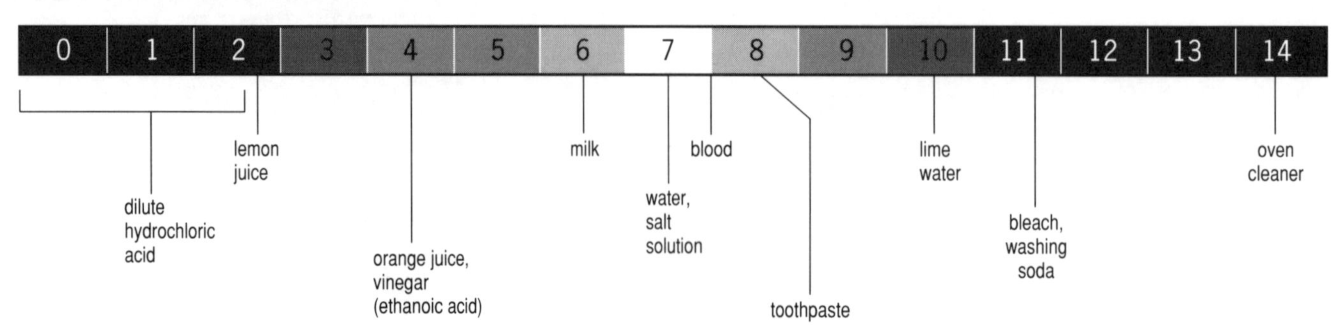

Figure 2.6.1 The pH scale

Strong and weak alkalis

An alkali is a substance which produces hydroxide ions ($OH^-(aq)$) when it is dissolved in water. Sodium hydroxide is a **strong alkali** because when it dissolves in water the ions separate completely.

$$\text{sodium hydroxide} \xrightarrow{\text{water}} \text{sodium ions} + \text{hydroxide ions}$$

$$NaOH(aq) \longrightarrow Na^+(aq) + OH^-(aq)$$

Substances which are strong alkalis have a high pH. A **weak alkali** such as ammonia produces only a few hydroxide ions when it dissolves in water:

$$\text{ammonia} + \text{water} \rightleftharpoons \text{ammonium ions} + \text{hydroxide ions}$$

$$NH_3(g) + H_2O(l) \rightleftharpoons NH_4^+(aq) + OH^-(aq)$$

The ammonia molecules react with the water molecules to form ammonium ions and hydroxide ions. However, only a few ammonia molecules do this so only a low concentration of hydroxide ions are produced.

Neutralising an acid

An alkali such as sodium hydroxide can be used to **neutralise** a common acid such as hydrochloric acid. If the pH of the acid is measured as sodium hydroxide solution is added to it, then the pH increases. If equal volumes of the same concentration of hydrochloric acid and sodium hydroxide are added to one another, the resulting solution is found to have a pH of 7. The acid has been neutralised and a neutral solution has been formed.

$$\text{hydrochloric acid} + \text{sodium hydroxide} \longrightarrow \text{sodium chloride} + \text{water}$$

$$HCl(aq) + NaOH(aq) \longrightarrow NaCl(aq) + H_2O(l)$$

When hydrochloric acid and sodium hydroxide dissolve in water the ions separate completely. We may therefore write:

$$H^+(aq) + Cl^-(aq) + Na^+(aq) + OH^-(aq) \longrightarrow Na^+(aq) + Cl^-(aq) + H_2O(l)$$

You will notice that certain ions are unchanged on either side of the equation. They are called **spectator ions** and are usually taken out of the equation. The equation now becomes:

$$H^+(aq) + OH^-(aq) \longrightarrow H_2O(l)$$

This type of equation is known as an **ionic equation**. The reaction between any acid and alkali in aqueous solution can be summarised by this ionic equation. It shows the ion which causes acidity ($H^+(aq)$) reacting with the ion which causes alkalinity ($OH^-(aq)$) to produce neutral water ($H_2O(l)$).

Quick Questions

1 Explain what is understood by the terms:
   a) Strong acid;
   b) Weak acid;
   c) Strong alkali;
   d) Weak alkali;
   e) Concentrated acid;
   f) Neutralisation.

2 Alongside the names of various chemicals are shown the pH values of that chemical in aqueous solution.

Potassium hydroxide	pH 13
Hydrogen bromide	pH 2
Calcium hydroxide	pH 11
Sodium chloride	pH 7
Hydrogen chloride	pH 2
Magnesium hydroxide	pH 10
Citric acid	pH 4

   Which of these substances is/are:
   a) A strong acid?
   b) A weak acid?
   c) A strong alkali?
   d) A weak alkali?
   e) A neutral substance?
   In each case write a chemical equation to show the molecules/ions present in solution.

3 Write a chemical equation to represent the neutralisation of sulphuric acid by sodium hydroxide. Reduce this to the ionic equation. Account for any difference you see between the ionic equation you have written and that shown for the reaction between hydrochloric acid and sodium hydroxide shown above.

2.7 Soluble salts

A **salt** is a compound which has been formed when all the hydrogen atoms of an acid have been replaced by metal atoms or by the ammonium ion (NH_4^+). Salts are very useful substances.

Table 2.7.1

Salt	Use
Silver bromide	In photography
Calcium carbonate	Extraction of iron, making cement, glass making
Sodium carbonate	Glass making, softening water, making modern washing powders
Ammonium chloride	In torch batteries
Calcium chloride	In the extraction of sodium, drying agent (anhydrous)
Sodium chloride	Making hydrochloric acid via the chlor-alkali industry, for food flavouring, hospital saline, in the solvay process for the manufacture of sodium carbonate
Tin(II) fluoride	Additive to toothpaste
Potassium nitrate	In fertiliser and gunpowder manufacture
Sodium stearate	In soap manufacture
Ammonium sulphate	In fertilisers
Calcium sulphate	For making plaster boards, plaster casts for injured limbs
Iron(II) sulphate	In 'iron' tablets
Magnesium sulphate	In medicines

If the acid being neutralised is hydrochloric acid, then salts called chlorides are formed. Other types of salts can be formed with other acids (see Table 2.7.2).

Table 2.7.2

Acid	Type of salt	Example
Carbonic acid	Carbonates	Sodium carbonate (Na_2CO_3)
Ethanoic acid	Ethanoates	Sodium ethanoate (CH_3COONa)
Hydrochloric acid	Chlorides	Potassium chloride (KCl)
Nitric acid	Nitrates	Potassium nitrate (KNO_3)
Sulphuric acid	Sulphates	Sodium sulphate (Na_2SO_4)

Methods of preparing soluble salts

There are four general methods of preparing soluble salts.

1 Acid + metal

This method can only be used with the less reactive metals. The metals usually used are the 'MAZIT' metals, that is, magnesium, aluminium, zinc, iron and tin.

Excess magnesium ribbon, for example, is added to dilute hydrochloric acid. Effervescence is observed due to the production of hydrogen gas.

$$\text{magnesium} + \text{hydrochloric acid} \longrightarrow \text{magnesium chloride} + \text{hydrogen}$$

$$Mg(s) + 2HCl(aq) \longrightarrow MgCl_2(aq) + H_2(g)$$

The excess magnesium is removed by filtration. The magnesium chloride solution is evaporated slowly to form a saturated solution of the salt.

The hot concentrated magnesium chloride solution so produced is tested by dipping a cold glass rod into it. If salt crystals form on the end of the rod, the solution is ready to crystallise and so is left to cool. Any crystals produced are filtered and dried between clean tissues.

2 Acid + carbonate

This method can be used with any metal carbonate and any acid, providing the salt produced is soluble. The typical experimental procedure is similar to that carried out between an acid and a metal. For example, copper(II) carbonate is added in excess to dilute nitric acid. Effervescence is observed due to the production of carbon dioxide.

$$\text{copper(II) carbonate} + \text{nitric acid} \longrightarrow \text{copper(II) nitrate} + \text{carbon dioxide} + \text{water}$$

$$CuCO_3(s) + 2HNO_3(aq) \longrightarrow Cu(NO_3)_2(aq) + CO_2(g) + H_2O(l)$$

Metal carbonates contain carbonate ions, CO_3^{2-}. It is the carbonate ions, which react with the hydrogen ions in the acid, that are responsible for the reaction which occurs.

$$\text{carbonate ions} + \text{hydrogen ions} \longrightarrow \text{carbon dioxide} + \text{water}$$

$$CO_3^{2-}(aq) + 2H^+(aq) \longrightarrow CO_2(g) + H_2O(l)$$

3 Acid + alkali

This method is generally used for preparing the salts of very reactive metals such as potassium or sodium. It would be too dangerous to add the metal directly to the acid, so instead we use an alkali which contains the reactive metal.

A **base** is a substance which neutralises an acid, producing a salt and water as the only products. In general, most metal oxides and hydroxides (as well as ammonia solution) are bases. If the base is soluble the term alkali can be used, but there are several bases which are insoluble. Salts can only be formed by this method if the base is soluble.

Table 2.7.3

Soluble bases (alkalis)	Insoluble bases
Sodium hydroxide (NaOH)	Iron(III) oxide (Fe_2O_3)
Potassium hydroxide (KOH)	Copper(II) oxide (CuO)
Calcium hydroxide ($Ca(OH)_2$)	Lead(II) oxide (PbO)
Ammonia solution ($NH_3(aq)$)	Magnesium oxide (MgO)

The preparation is carried out by titration, as described in Section 3.17. Once you know how much acid is required to reach the end-point, you can add the same volume of acid to the measured volume of alkali but this time without the indicator. The solution which is produced can then be evaporated slowly to obtain the salt.

For example:

hydrochloric acid + sodium hydroxide ⟶ sodium chloride + water

$$HCl(aq) + NaOH(aq) \longrightarrow NaCl(aq) + H_2O(l)$$

4 Acid+insoluble base

This method can be used to prepare a salt of unreactive metals such as lead and copper. In these cases, it is not possible to use a direct reaction of the metal with an acid so instead the acid is neutralised using the particular metal oxide. The method is generally the same as that carried out between a metal carbonate and an acid, though some warming of the reactants may be necessary. An example of such a reaction is the neutralisation of sulphuric acid by copper(II) oxide.

sulphuric acid + copper(II) oxide ⟶ copper(II) sulphate + water

$$H_2SO_4(aq) + CuO(s) \longrightarrow CuSO_4(aq) + H_2O(l)$$

Metal oxides contain the oxide ion, O^{2-}. The ionic equation for this reaction is therefore:

$$2H^+(aq) + O^{2-}(s) \longrightarrow H_2O(l)$$

The copper sulphate produced by this method is in the form of a **salt hydrate**. A hydrate is a salt which incorporates water molecules into its crystal structure. This water is referred to as **water of crystallisation**. In this example, five water molecules are incorporated into the structure. The formula for hydrated copper(II) sulphate is $CuSO_4.5H_2O$.

It is for this reason that the crystals are formed by slow evaporation, followed by crystallisation from a saturated solution. If copper(II) sulphate solution was heated strongly until no water was left then **anhydrous** copper(II) sulphate would be produced - formula $CuSO_4$.

Quick Questions

1 Complete the word equations and write balanced chemical equations for the following soluble salt preparations:

a) Magnesium + sulphuric acid ⟶

b) Calcium carbonate + hydrochloric acid ⟶

c) Zinc oxide + hydrochloric acid ⟶

d) Potassium hydroxide + nitric acid ⟶

Write ionic equations for each of the reactions.

2 a) Copy out and complete the table which is about the different methods of preparing salts.

Method of preparation	Name of salt prepared	Two substances used in the preparation
Acid + alkali	Sodium sulphate	________ and ________
Acid + metal	________	________ and dilute hydrochloric acid
Acid + insoluble base	Copper(II) sulphate	________ and ________
Acid + carbonate	Magnesium ________	________ and ________

b) Write word and balanced chemical equations for each reaction shown in your table.

2.8 Solubility

Solubility of salts in water

Water is a very good solvent and will dissolve a whole range of solutes. You can dissolve more sugar than sodium chloride in 100 cm^3 of water at the same temperature. The sugar is more soluble than the sodium chloride at the same temperature. The **solubility** of a solute in water at a given temperature is the number of grams of that solute which can be dissolved in 100 g of water to produce a saturated solution at that temperature.

Calculating solubility

Example.
21.5 g of sodium chloride dissolves in 60 g of water at 25 °C. Calculate the solubility of sodium chloride in water at that temperature.

If 60 g of water dissolves 21.5 g of sodium chloride then 1 g of water will dissolve:

$$\frac{21.5}{60} \text{ g of sodium chloride}$$

Therefore, 100 g of water will dissolve:

$$\frac{21.5}{60} \times 100 = 35.8\text{g}$$

The solubility of sodium chloride at 25 °C =35.8 g/100 g of water.

The rate at which a solute dissolves depends on the temperature and volume of the solvent used, the surface area of the solute and the amount of stirring.

Solubility curves

Usually, the amount of solute that a solvent will dissolve increases with temperature. Figure 2.8.1 shows how the solubility of copper(II) sulphate and potassium nitrate vary with temperature. These graphs of solubility against temperature are known as **solubility curves**.

From the solubility curve in Figure 2.8.1, it can be seen that the solubility of copper(II) sulphate at 30 °C is 22 g whilst that of potassium nitrate is 44 g. Also, when a saturated solution is cooled some of the solute crystallises out of solution. Therefore, by using solubility curves it is possible to determine the amounts of solute that will crystallise out of solution at different temperatures. For example, if potassium nitrate is cooled from 70 °C to 40 °C, then the difference between the two solubilities at these temperatures tells you the amount that crystallises out.

Solubility at 70 °C	=	135 g
Solubility at 40 °C	=	62 g
The difference	=	73 g

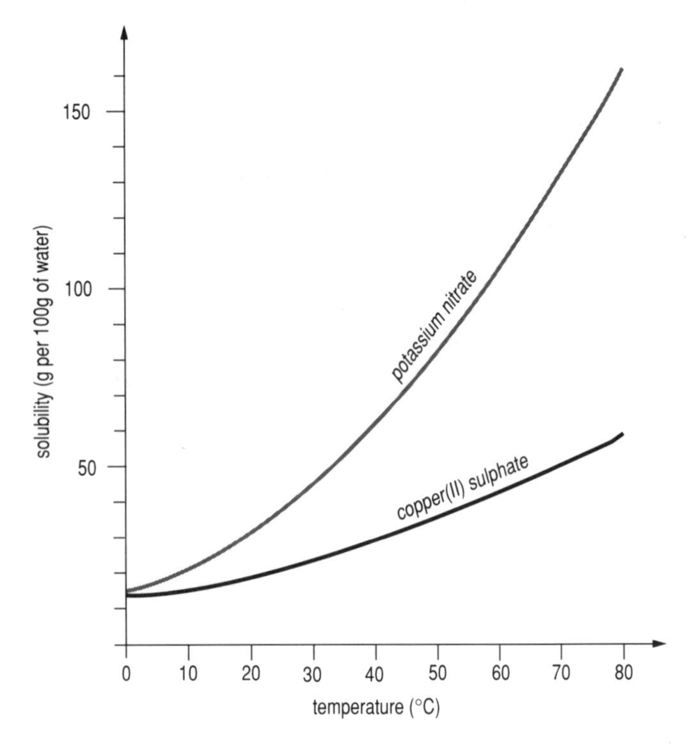

Figure 2.8.1 Solubility curves for copper(II) sulphate and potassium nitrate

Solubility of gases

Gases become less soluble at higher temperatures – the reverse of solubility in solids. As temperatures rise, gas particles have sufficient energy to escape from the solvent in which they are dissolved.

Rules for solubility

Soluble salts can be classified into those that are soluble or insoluble in water. The following salts are soluble in cold water:

- all nitrates.
- all common sodium, potassium and ammonium salts.
- all chlorides except lead, silver and mercury.
- all sulphates except lead, barium and calcium.

Methods of preparing insoluble salts

An insoluble salt such as barium sulphate can be made by **precipitation**. In this case, solutions of the two chosen soluble salts are mixed. To produce barium sulphate, barium nitrate and sodium sulphate can be used. The barium sulphate precipitate can be filtered off, washed with distilled water and dried. Barium sulphate is used in X-ray identification of stomach ulcers. The reaction that has occurred is:

$$\text{barium nitrate} + \text{sodium sulphate} \longrightarrow \text{barium sulphate} + \text{sodium nitrate}$$

$$Ba(NO_3)_2(aq) + Na_2SO_4(aq) \longrightarrow BaSO_4(s) + NaNO_3(aq)$$

The ionic equation for this reaction is:

$$Ba^{2+}(aq) + SO_4^{2-}(aq) \longrightarrow BaSO_4(s)$$

This method is sometimes known as **double decomposition** and may be summarised as:

$$\text{soluble salt} + \text{soluble salt} \longrightarrow \text{insoluble salt} + \text{soluble salt}$$

It should be noted that even salts like barium sulphate dissolve to a very small extent. This substance and others like it are said to be sparingly soluble.

Testing for different salts

There are simple chemical tests which allow us to identify the anion part of the salt.

Testing for a sulphate

If you take a solution of a suspected soluble sulphate and add to it a few drops of dilute hydrochloric acid and a soluble barium salt (e.g. barium chloride) then a white precipitate of barium sulphate is produced.

$$\text{barium ion} + \text{sulphate ion} \longrightarrow \text{barium sulphate}$$
$$Ba^{2+}(aq) + SO_4^{2-}(aq) \longrightarrow BaSO_4(s)$$

Testing for a chloride

If you take a solution of the suspected soluble chloride and add to it a few drops of dilute nitric acid and then a small volume of a soluble silver salt (e.g. silver nitrate) then a white precipitate of silver chloride is produced.

$$\text{silver ion} + \text{chloride ion} \longrightarrow \text{silver chloride}$$
$$Ag^{+}(aq) + Cl^{-}(aq) \longrightarrow AgCl(s)$$

If left to stand the precipitate goes grey. A similar test can be carried out to test for bromides and iodides. Bromides give a cream precipitate of silver bromide and iodides a yellow precipitate of silver iodide.

Testing for a carbonate

If a small amount of an acid is added to some of the suspected carbonate, then effervescence occurs. If it is a carbonate then carbon dioxide gas is produced. This will turn limewater milky (see Section 3.4).

$$\text{carbonate ions} + \text{hydrogen ions} \longrightarrow \text{carbon dioxide} + \text{water}$$
$$CO_3^{2-}(aq) + 2H^{+}(aq) \longrightarrow CO_2(g) + H_2O(l)$$

Quick Questions

1 Calculate the solubility, at the temperature given, of the following salts in water:

a) 94.5 g potassium nitrate dissolves in 70 g of water at 68 °C.

b) 9.2 g of ammonium chloride dissolves in 20 g of water at 40 °C.

2 Use the data given below to plot a solubility curve of ammonium chloride.

Temp/°C	0	10	20	30	40	50	60	70	80
Solubility g/100 g of water	29.2	33.0	37.1	41.8	45.8	50.8	55.2	60.2	65.6

a) Using your solubility curve find the solubility of ammonium chloride at:
i) 15 °C
ii) 35 °C,
iii) 65 °C.

b) Using your answers to part **a)**, calculate the amount of ammonium chloride which will crystallise out of solution when cooled from:
i) 65 °C to 35 °C,
ii) 35 °C to 15 °C.

c) Calculate the amount of ammonium chloride which would crystallise out of solution if 25 g of the saturated solution were cooled from 52 °C to 23 °C.

3 An analytical chemist working for an environmental health organisation has been given a sample of water which is thought to have been contaminated with a sulphate and a chloride.

a) Describe how she could confirm the presence of these two types of salt by simple chemical tests.

b) Write ionic equations to help you explain what is happening during the testing process.

2.9 Water

In some hot, arid countries the sea is the main source of pure drinking water. The water is obtained by a process known as **desalination**.

Sea water contains about 35 g of dissolved solids in each kilogram. A typical analysis of the elements present in sea water is shown in Table 2.9.1.

Table 2.9.1

Element	g/dm^3 of sea water
Bromine	0.07
Calcium	0.4
Chlorine	19.2
Magnesium	1.3
Potassium	0.4
Sodium	10.7
Sulphur	0.9
Other elements	1.4

The elements shown above are present as constituents of compounds, and not as the free element. Some elements and compounds are extracted from sea water on a commercial basis - for example bromine, magnesium and sodium chloride.

Sodium chloride

This is the most abundant resource in sea water. There is about 25 g of sodium chloride in every cubic decimetre of sea water. It is extracted in several areas of the world by evaporation, for example, in France, Saudi Arabia and Australia.

The sea water is kept in shallow ponds until all the water has been evaporated by the heat of the Sun. The salt is then harvested.

Sodium chloride is used to flavour food, in the manufacture of sodium carbonate and sodium hydrogen carbonate and it is the raw material for the **chlor-alkali industry**.

The water cycle

The diagram of the **water cycle** below shows how water circulates around the Earth. The driving force behind the water cycle is the heat of the Sun which causes evaporation from oceans, seas and lakes. Water vapour is also formed from the evaporation of water from leaves (transpiration), through **respiration** and **combustion**. The water vapour rises and cools, condensing to form tiny droplets of water. These droplets form clouds.

Clouds are moved along by air currents. As they cool, the tiny droplets join to form larger droplets. These fall as rain when the water droplets reach a certain size.

The water which falls as rain runs into streams and rivers and then on into lakes, seas and oceans.

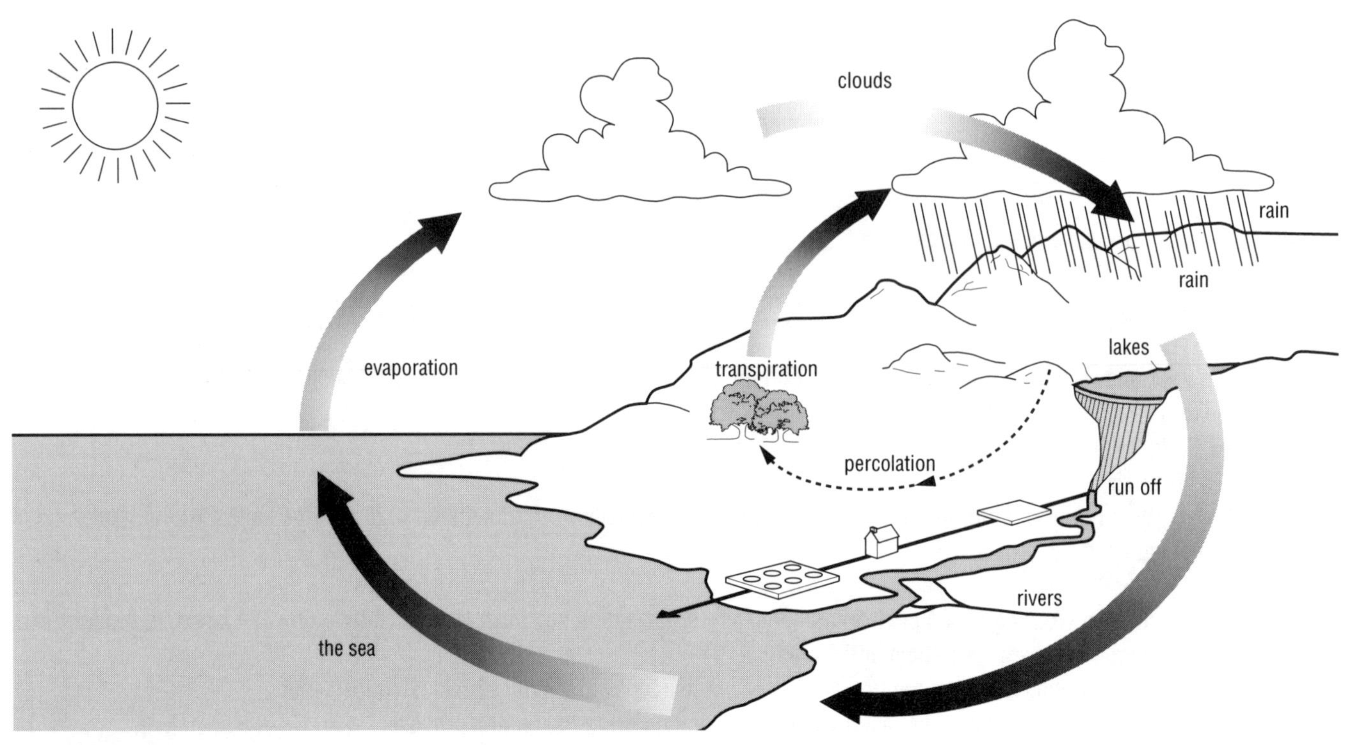

Figure 2.9.1 The water cycle

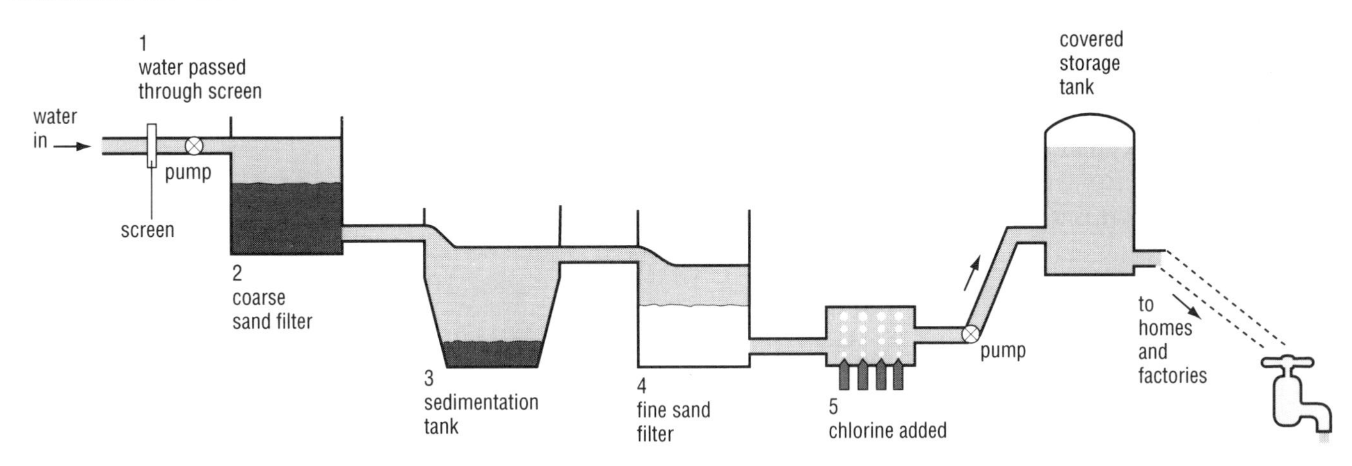

1 Impure water is first passed through screens which filter out floating debris.

2 Filtration through coarse sand traps larger, insoluble particles. The sand also contains specially grown microbes which remove some of the bacteria.

3 A sedimentation tank has chemicals known as flocculants added to it to make the smaller particles, which remain in the water, stick together. They sink to the bottom of the tank.

4 These particles are then removed by further filtration through fine sand.

5 Finally a little chlorine gas is added. It kills any remaining bacteria and sterilises the water.

Figure 2.9.2 The processes involved in water treatment

Water pollution

Water is very good at dissolving substances. It is, therefore, very unusual to find really pure water on this planet. As water falls through the atmosphere and down onto and through the surface of the earth, it dissolves a tremendous variety of substances. For example, river water will contain grit, dust and bacteria. Chemical fertilisers washed off surrounding land will add nitrate ions (NO_3^-) and phosphate ions (PO_4^{3-}) to the water. It may also contain human waste as well as insoluble impurities such as oil and lead dust (to a decreasing extent) from the exhaust fumes of lorries and cars. These impurities must be removed from the water before it can be used - see Figure 2.9.2. Most drinking water in the United Kingdom is obtained from lakes and rivers where the pollution level is low.

Sewage treatment

After we have used water it must again be treated before it can be returned to rivers, lakes and seas. The process known as sewage treatment is shown in the diagram alongside (Figure 2.9.3).

Used water, sewage, contains waste products such as human waste and washing up water as well as everything else that we put down a drain or a sink. The processes involved in treatment are:

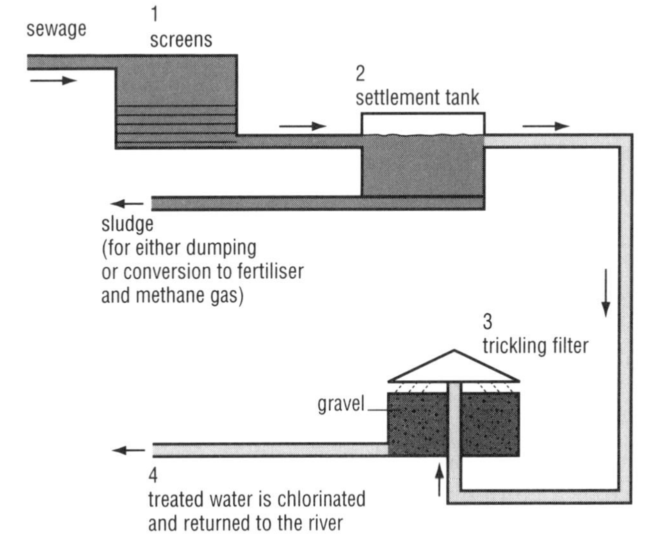

1 Large screens are used to remove large pieces of rubbish.
2 Sand and grit are separated in large sedimentation tanks. The sand and grit often contain large amounts of useful chemicals which, by the action of selected microbes, can be used as fertilisers.
3 The impure water is then removed and sent to a trickling filter where it is allowed to drain through gravel on which microbes have been deposited. These kill off the bacteria that still remain in the water.
4 Finally the treated water is chlorinated and returned to a river.

Figure 2.9.3 The processes involved in sewage treatment

Chemical tests for water

The presence of water in a substance can be tested using anhydrous copper(II) sulphate which turns from white to blue if water is present, or anhydrous cobalt(II) chloride which turns from blue to pink. These tests do not indicate that a substance is pure water - only that water is present.

Quick Questions

1 Construct a simplified version of the water cycle using 'key-words' in boxes and the 'processes involved' over linking arrows.
2 Make a list of four major water pollutants and explain where they come from. What damage can these pollutants do?

2.10 Hardness of water

Rain water dissolves carbon dioxide as it falls through the atmosphere. A small fraction of this dissolved carbon dioxide reacts with the water to produce carbonic acid, and it is a weak acid.

$$\text{water} + \text{carbon dioxide} \rightleftharpoons \text{carbonic acid}$$
$$H_2O(l) + CO_2(g) \rightleftharpoons H_2CO_3(aq)$$

As this solution passes over and through rocks containing limestone and dolomite, the weak acid attacks these rocks and very slowly dissolves them. The dissolved substances are called calcium and magnesium hydrogencarbonates.

$$\text{calcium carbonate} + \text{carbonic acid} \longrightarrow \text{calcium hydrogencarbonate}$$
$$CaCO_3(s) + H_2CO_3(aq) \longrightarrow Ca(HCO_3)_2(aq)$$

Some of the rock strata may also contain calcium and magnesium sulphates which are only very sparingly soluble in water. The presence of these dissolved substances causes the water to become **hard**. Hardness in water can be divided into two types - **temporary** and **permanent hardness**. Temporary hardness is caused by the presence of dissolved calcium or magnesium hydrogencarbonates. It can be removed easily by boiling. Permanent hardness is caused by the presence of dissolved calcium or magnesium sulphates and cannot be removed completely by boiling.

When water containing these substances is boiled or evaporated, a white solid deposit is left behind. These white deposits are calcium or magnesium sulphates and/or calcium carbonate.

It is this calcium carbonate which causes the furring in kettles and blockages in hot water pipes which occur in hard water areas.

Stalactites and stalagmites are found in underground caverns in limestone areas. They are formed from the slow decomposition of dissolved calcium or magnesium hydrogencarbonates.

$$\text{calcium hydrogencarbonate} \xrightarrow{\text{heat}} \text{calcium carbonate} + \text{water} + \text{carbon dioxide}$$
$$Ca(HCO_3)_2(aq) \longrightarrow CaCO_3(s) + H_2O(l) + CO_2(g)$$

Effect of hard water on soap

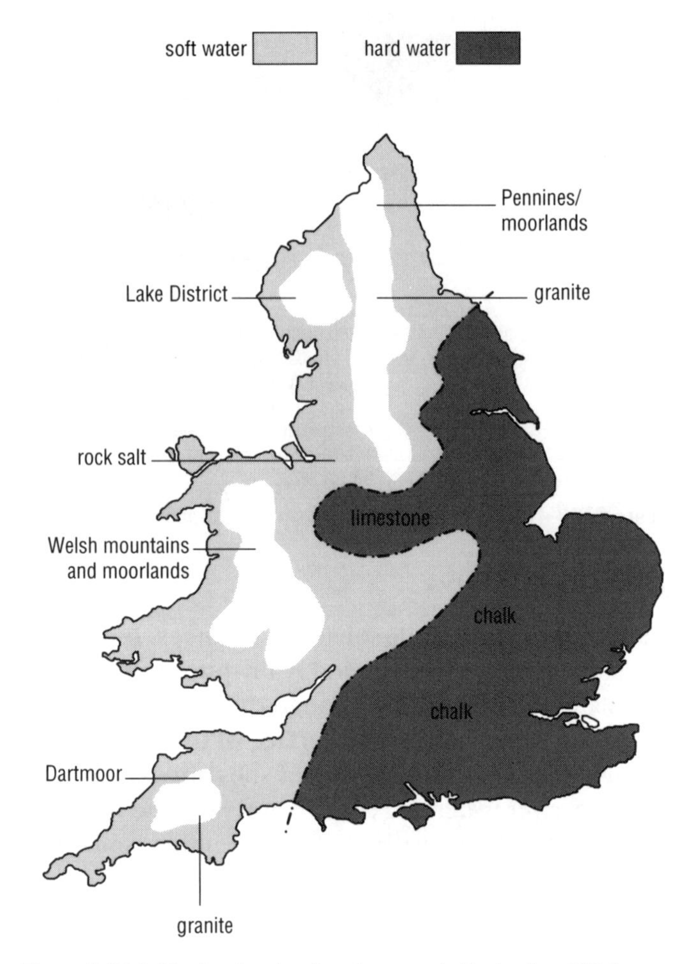

Figure 2.10.1 The hard and soft water areas in England and Wales

If you live in a hard water area, you may have noticed that it can be difficult to make your soap lather. Instead, the water becomes cloudy. This cloudiness is caused by the presence of a solid material (a precipitate), formed by the reaction of the dissolved substances in the water with soap (sodium stearate). This white precipitate is known as scum.

$$\text{sodium stearate (soap)} + \text{calcium hydrogencarbonate} \longrightarrow \text{calcium stearate} + \text{sodium hydrogencarbonate}$$
$$2\,NaSt(aq) + Ca(HCO_3)_2(aq) \longrightarrow Ca(St)_2(s) + 2NaHCO_3(aq)$$

$$St = \text{stearate} \quad NaSt = C_{17}H_{35}COO^-Na^+$$

The amount of soap required to just produce a lather can be used to estimate the hardness of the water.

To overcome the problem of scum formation, soapless detergents have been manufactured that do not produce a scum in hard water because they do not react with the substances dissolved in hard water.

Removal of hardness

Temporary hardness is easily removed from water by boiling. When heated, the calcium hydrogencarbonate decomposes to produce calcium carbonate which is insoluble.

$$\text{calcium hydrogencarbonate} \xrightarrow{\text{heat}} \text{calcium carbonate} + \text{water} + \text{carbon dioxide}$$

$$Ca(HCO_3)_2(aq) \longrightarrow CaCO_3(s) + H_2O(l) + CO_2(g)$$

The substances in permanently hard water are not decomposed when heated and therefore cannot be removed by boiling. Both types of hardness can be removed by one of the following three methods.

Addition of washing soda ($Na_2CO_3.10H_2O$) crystals

In each case, the calcium or magnesium ion which actually causes the hardness is removed as a precipitate and can, therefore, no longer cause hardness.

$$\text{calcium ions (from hard water)} + \text{carbonate ions (from washing soda)} \longrightarrow \text{calcium carbonate}$$

$$Ca^{2+}(aq) + CO_3^{2-}(aq) \longrightarrow CaCO_3(s)$$

Ion exchange

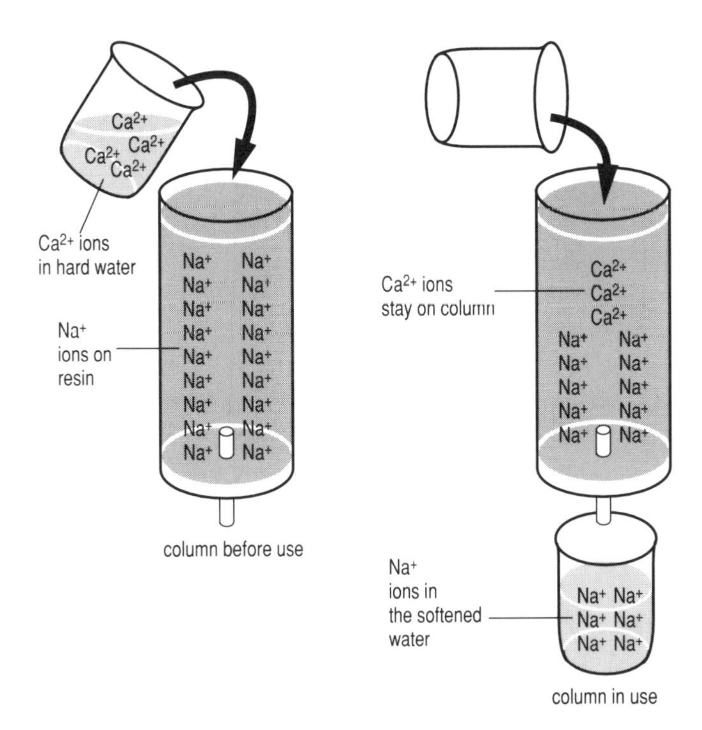

Figure 2.10.2 The ion exchange process

The water is passed through a container filled with a suitable resin containing sodium ions. The calcium or magnesium ion causing the hardness is exchanged for the sodium ions in the resin.

$$\text{calcium ion} + \text{sodium–resin} \longrightarrow \text{calcium–resin} + \text{sodium ion}$$

$$Ca^{2+}(aq) + Na_2^{+}-R^{2-}(s) \longrightarrow Ca^{2+}-R^{2-}(s) + 2Na^{+}(aq)$$

When all the sodium ions have been removed, the resin can be regenerated by pouring a solution of a suitable sodium salt through it (usually sodium chloride solution).

Distillation

Here the water is distilled away from the dissolved substances. This method is far too expensive to be used on a large scale.

Advantages and disadvantages of hard water

Table 2.10.1

Disadvantages of hard water	Advantages of hard water
Wastes soap.	Coats lead pipes with a thin layer of lead sulphate or lead carbonate and cuts down the possibility of lead poisoning.
Causes kettles to fur.	Calcium ions in hard water are required by the body for teeth and bones.
Can cause hot water pipes to block.	Has a nice taste.
Can spoil the finish of some fabrics.	Is good for brewing beer.

Quick Questions

1 One of the substances found in some temporary hard waters is magnesium hydrogencarbonate. Write a word and balanced chemical equation to show the effect of heat on this substance in aqueous solution.

2 Explain the following:

a) Industry normally requires water that has been softened;

b) Hard water causes kettles to fur;

c) Hard water wastes soap.

2.11 Electrolysis of solutions

Electrolysis of water

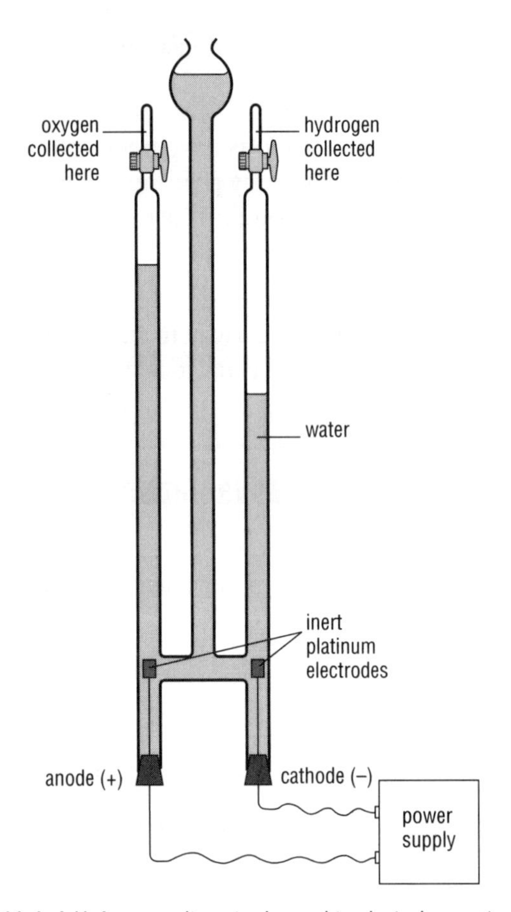

Figure 2.11.1 A Hofmann voltameter is used to electrolyse water

Pure water is a very poor conductor of electricity because there are so few ions in it. However, it can be made to decompose if an electric current is passed through it in a Hofmann voltameter.

To enable water to conduct electricity better, some dilute sulphuric acid (or sodium hydroxide solution) is added. When an electric current flows through this solution, gases can be seen to be produced at the two electrodes. After about twenty minutes, roughly twice as much gas has been produced at the cathode as at the anode. The gas collected at the cathode burns with a squeaky 'pop', showing it to be hydrogen gas.

Positively charged hydrogen ions must have been attracted to the cathode.

hydrogen ions + electrons ⟶ hydrogen molecules

$$4H^+(aq) + 4e^- \longrightarrow 2H_2(g)$$

If, during this process, the water molecules lose $H^+(aq)$ then the remaining portion must be hydroxide ions, $OH^-(aq)$. These ions are attracted to the anode. The gas collected at the anode relights a glowing spill, showing it to be oxygen.

This gas is produced in the following way.

hydroxide ions ⟶ water molecules + oxygen molecules + electrons

$$4OH^-(aq) \longrightarrow 2H_2O(l) + O_2(g) + 4e^-$$

This experiment was first carried out by Sir Humphry Davy who confirmed that the formula for water, by this experiment, was H_2O.

Purification of copper

Because copper is a very good conductor of electricity it is used for electrical wiring and cables. However, even small amounts of impurities cut down this conductivity quite noticeably. The metal must be 99.99% pure to be used in this way. To ensure this level of purity, the newly extracted copper has to be purified by electrolysis. The impure copper is used as the anode. The cathode is made from very pure copper. The electrolyte is a solution of copper(II) sulphate acidified with sulphuric acid, added to improve conductivity.

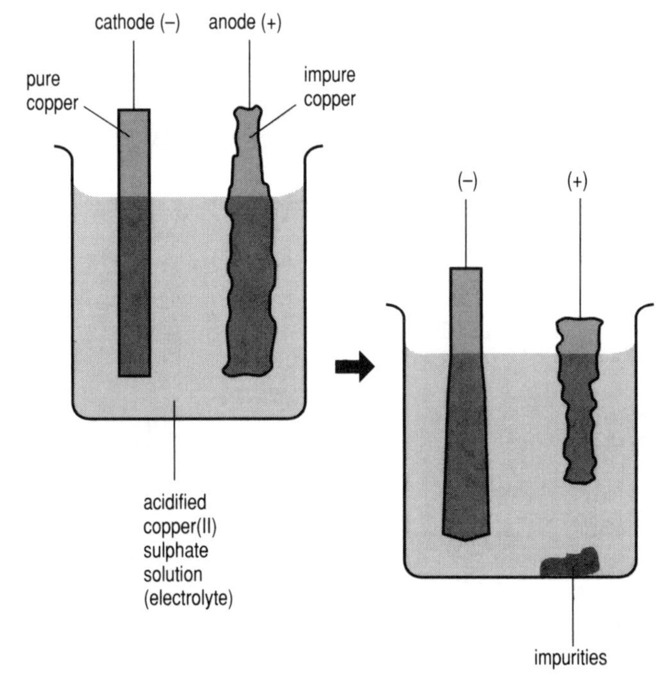

Figure 2.11.2 Copper purification process

When the current flows, the copper moves from the impure anode to the pure cathode. Any impurities fall to the bottom of the cell and collect below the anode in the form of a slime. This slime is rich in precious metals and the recovery of these metals is an important aspect of the economics of the process. The electrolysis proceeds for about three weeks until the anodes are reduced to about 10% of their original size.

The ions present in the solution are:
From the water: $H^+(aq)$, $OH^-(aq)$
From the copper(II) sulphate: $Cu^{2+}(aq)$, $SO_4^{2-}(aq)$

During the process, the impure anode loses mass because the copper atoms leave and lose electrons to become copper ions, $Cu^{2+}(aq)$, see Figure 2.11.3.

copper atoms ⟶ copper ions + electrons

$$Cu(s) \longrightarrow Cu^{2+}(aq) + 2e^-$$

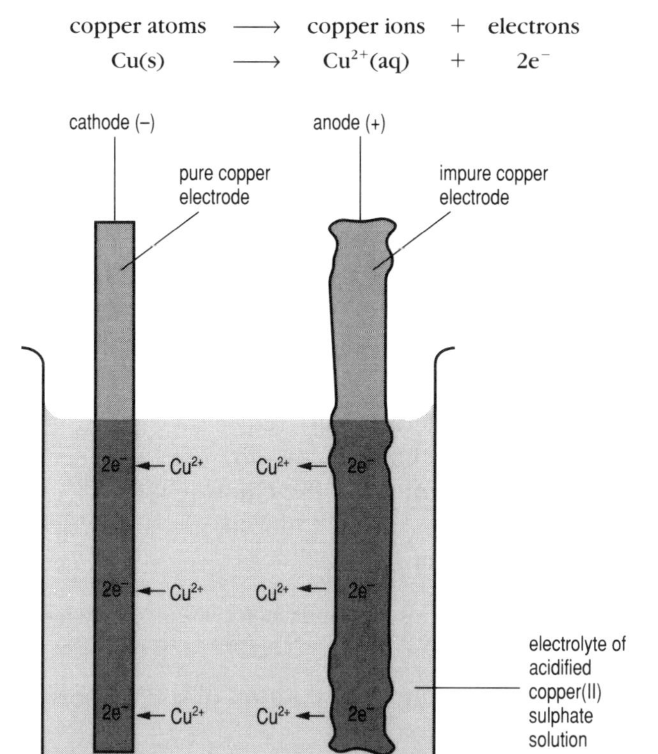

Figure 2.11.3 The movement of ions in the purification of copper by electrolysis

The electrons which are released at the anode travel around the external circuit to the cathode. There the electrons are passed onto the copper ions, $Cu^{2+}(aq)$, from the copper(II) sulphate solution and the copper is deposited on the cathode.

copper ions + electrons ⟶ copper atoms

$$Cu^{2+}(aq) + 2e^- \longrightarrow Cu(s)$$

Electrolysis of other solutions

Using inert electrodes, the following solutions undergo electrolysis to produce these products.

Table 2.11.1

Solution	Anode	Cathode
Hydrochloric acid	Chlorine	Hydrogen
Copper(II) chloride	Chlorine	Copper
Copper(II) sulphate	Oxygen	Copper

Electroplating of steel

Electroplating is the process which involves using electrolysis to plate, or coat, one metal with another. Often the purpose of electroplating is to give a protective coating to the metal beneath. For example, bath taps are chromium plated to prevent **corrosion**, but at the same time it gives them a more attractive finish.

The electroplating process is carried out in a cell such as the one shown in Figure 2.11.4. This is usually known as the 'plating bath' and it contains a suitable electrolyte.

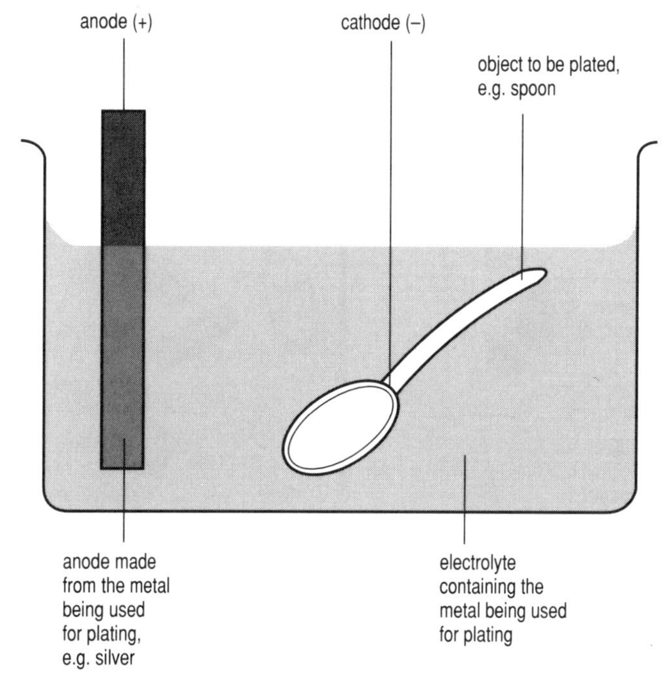

Figure 2.11.4 Silver plating a spoon

For silver plating, the electrolyte is a solution of a silver salt. The article to be plated is made the cathode in the cell so that the metal ions move to it when the current is switched on. The cathode reaction in this process is:

silver ions + electrons ⟶ silver atoms

$$Ag^+(aq) + e^- \longrightarrow Ag(s)$$

Quick Questions

1. Why is solid copper(II) bromide not an electrical conductor?
2. Why do ions move towards the different electrodes?
3. Suggest a reason for only 'roughly' twice as much hydrogen gas being produced at the cathode as oxygen gas at the anode.

2.12 Chlor-alkali industry

The electrolysis of saturated sodium chloride solution (brine) is the basis of a major industry. Three very important substances are produced - namely chlorine, sodium hydroxide and hydrogen.

The electrolytic process is a very expensive one, requiring vast amounts of electricity. The process is economic only because all three products have a large number of uses.

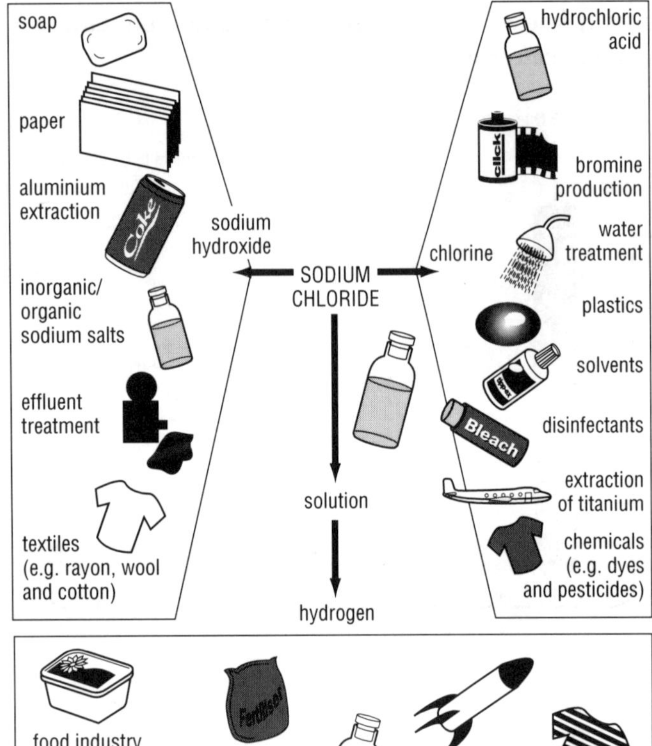

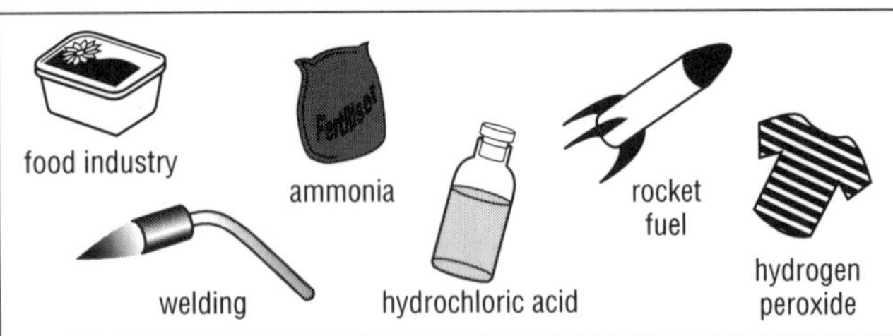

Figure 2.12.1 The use of substances from the chlor-alkali industry

Recent developments in electrolysis technology have developed the **membrane cell** (see Figure 2.12.2). This method is now preferred to other older methods (diaphragm and mercury cells) as it produces a purer product, it causes little pollution and is cheaper to run.

The extraction of sodium chloride for this process is usually carried out by solution mining which involves the salt being dissolved in water, before being brought to the surface. This type of mining can lead to problems with subsidence.

The brine is first purified to remove calcium, strontium and magnesium compounds by a process of ion exchange.

The membrane cell is used continuously, with fresh brine flowing into the cell as the process breaks the brine up into the products. The cell has been designed to ensure that the products do not mix.

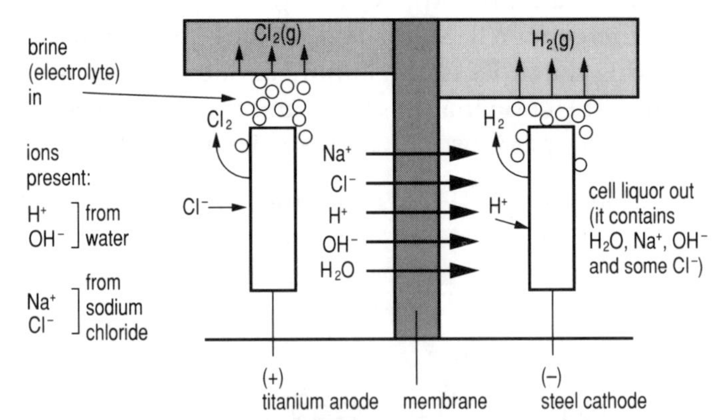

Figure 2.12.2 Cross-section through one compartment of a membrane cell

The ions in this concentrated sodium chloride solution are:
From the water: $H^+(aq)$ $OH^-(aq)$
From the sodium chloride: $Na^+(aq)$ $Cl^-(aq)$

When the current flows the chloride ions, $Cl^-(aq)$, are attracted to the anode. Chlorine gas is produced by the electrolysis process:

$$\text{chloride ions} \longrightarrow \text{chlorine molecules} + \text{electrons}$$
$$2Cl^-(aq) \longrightarrow Cl_2(g) + 2e^-$$

This leaves a high concentration of sodium ions, $Na^+(aq)$, around the anode.

The hydrogen ions, $H^+(aq)$, ions are attracted to the cathode and hydrogen gas is produced.

$$\text{hydrogen ions} + \text{electrons} \longrightarrow \text{hydrogen molecules}$$
$$2H^+(aq) + 2e^- \longrightarrow H_2(g)$$

This leaves a high concentration of hydroxide ions, $OH^-(aq)$, around the cathode. The sodium ions, $Na^+(aq)$, are drawn through the membrane where they combine with the $OH^-(aq)$ ions to form sodium hydroxide, NaOH, solution.

Reactions of sodium hydroxide

Sodium hydroxide is the most commonly used of the strong alkalis. It is commonly known as caustic soda, and is a corrosive substance.

If solid sodium hydroxide is left in the air it absorbs water vapour and eventually forms a very concentrated solution of sodium hydroxide. This process is called **deliquescence**.

As an alkali, it will neutralise an acid to give a salt and water.

$$\text{sodium hydroxide} + \text{hydrochloric acid} \longrightarrow \text{sodium chloride} + \text{water}$$
$$NaOH(aq) + HCl(aq) \longrightarrow NaCl(aq) + H_2O(l)$$

Sodium hydroxide readily reacts with solutions of metallic compounds to form hydroxide precipitates.

The colour and solubility of these precipitates can be used to identify metal ions in solution (see Section 2.3).

It is used to produce pure aluminium oxide from bauxite, from which aluminium metal is extracted by electrolysis. The manufacture of soaps and detergents also involves reactions with sodium hydroxide.

Bleach is made by dissolving chlorine gas in a cold dilute solution of sodium hydroxide.

$$\text{sodium hydroxide} + \text{chlorine} \longrightarrow \text{sodium chloride} + \text{sodium chlorate(I)} + \text{water}$$
$$2NaOH(aq) + Cl_2(g) \longrightarrow NaCl(aq) + NaClO(aq) + H_2O(l)$$

The solution formed from this reaction, that of a one to one ratio of sodium chloride and sodium chlorate(I), is commonly called household bleach.

When carbon dioxide is bubbled through a solution of carbon dioxide, sodium carbonate and water are formed.

$$\text{carbon dioxide} + \text{sodium hydroxide} \longrightarrow \text{sodium carbonate} + \text{water}$$
$$CO_2(g) + 2NaOH(aq) \longrightarrow Na_2CO_3(aq) + H_2O(l)$$

Preparation, properties and reactions of chlorine gas

Chlorine gas can be prepared in the laboratory by the oxidation of concentrated hydrochloric acid by potassium manganate(VII). Using this method no heat is required. The chlorine gas can be dried by passing it through concentrated sulphuric acid.

$$\text{hydrochloric acid} + \text{potassium manganate(VII) (oxidising agent)} \longrightarrow \text{chlorine} + \text{water}$$
$$2HCl(l) + [O] \longrightarrow Cl_2(g) + H_2O(l)$$

Chlorine gas has:

- A pale green colour at room temperature.
- A density greater than air.
- A pungent choking smell and is very poisonous.

Chlorine combines with many metals and non-metals such as iron and hydrogen. With iron, chlorine reacts to form iron(III) chloride.

$$\text{iron} + \text{chlorine} \longrightarrow \text{iron(III) chloride}$$
$$2Fe(s) + 3Cl_2(g) \longrightarrow 2FeCl_3(s)$$

A violent reaction can occur when hydrogen reacts with chlorine to form hydrogen chloride gas.

$$\text{hydrogen} + \text{chlorine} \longrightarrow \text{hydrogen chloride}$$
$$H_2(g) + Cl_2(g) \longrightarrow 2HCl(g)$$

In the above reaction, chlorine is acting as an oxidising agent. It shows this property more clearly when it is bubbled through a green solution of iron(II) ions. The solution gradually turns orange/brown as the iron(II) ions are oxidised to iron(III) ions.

Chlorine will react with alkalis to form chloride ions (Cl^-) and chlorate(I) ions (ClO^-). With cold sodium hydroxide solution it reacts to form sodium chloride and sodium chlorate(I), which together form a bleaching solution.

$$\text{sodium hydroxide} + \text{chlorine} \longrightarrow \text{sodium chloride} + \text{sodium chlorate(I)} + \text{water}$$
$$2NaOH(aq) + Cl_2(g) \longrightarrow NaCl(aq) + NaClO(aq) + H_2O(l)$$

Quick Questions

1. Account for the following observations that were made when concentrated sodium chloride solution, to which a little universal indicator had been added, was electrolysed in the laboratory.
   a) The universal indicator initially turns red in the region of the anode, but as the electrolysis proceeds loses its colour;
   b) The universal indicator turns blue in the region of the cathode.
2. Describe how you could use a dilute sodium hydroxide solution to distinguish between the following pairs of metal ions:
   a) Iron(II) ions and iron(III) ions;
   b) Aluminium ions and magnesium ions.
3. Write word and balanced chemical equations for the reaction between sodium hydroxide and sulphuric acid.

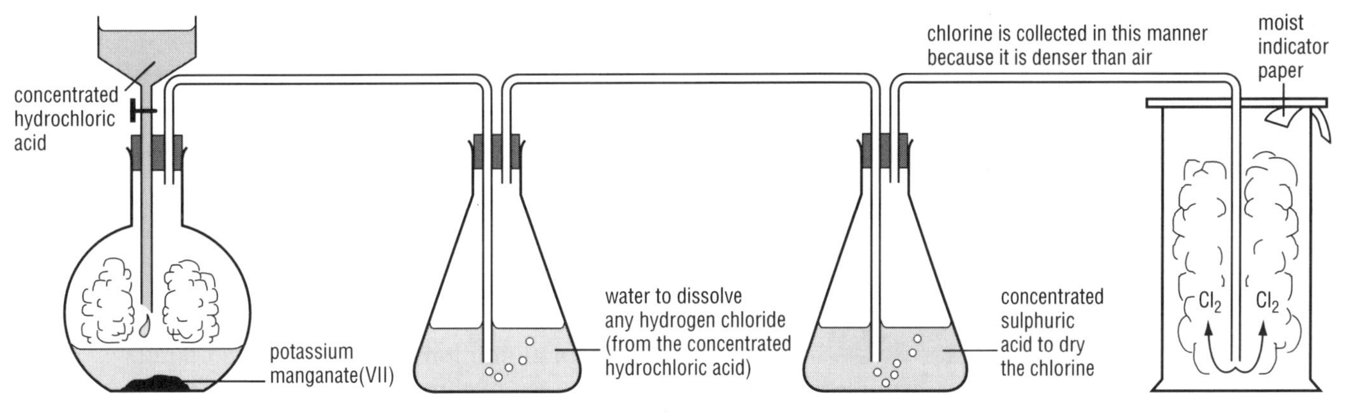

Figure 2.12.3

2.13 Calculations in electrolysis

The quantity of electricity flowing through an electrolysis cell is measured in coulombs (C). If one ampere is passed for one second, the quantity of electricity is said to be one coulomb.

$$\underset{\text{(coulombs)}}{\text{quantity of electricity}} = \underset{\text{(amps)}}{\text{current}} \times \underset{\text{(seconds)}}{\text{time}}$$

Therefore, if two amps flow for one second then the quantity of electricity passed is two coulombs.

In the purification of copper you saw that copper was deposited at the cathode. The electrode equation is:

$$\underset{\text{1 mole}}{Cu^{2+}(aq)} + \underset{\text{2 moles}}{2e^-} \longrightarrow \underset{\text{1 mole}}{Cu(s)}$$

This equation tells us that one mole of copper(II) ions combines with two moles of electrons to produce one mole of copper metal atoms (64 g).

A mole of electrons is called a **faraday**. This unit is named after an English scientist Michael Faraday (1791–1867) who carried out many significant experiments whilst investigating the nature of magnetism and electricity.

Figure 2.13.1 In 1883 Michael Faraday was the first scientist to measure the masses of elements produced during electrolysis

So we can say that we need two faradays of electricity to form one mole of copper atoms (64 g) at the cathode during this purification process. From accurate electrolysis experiments it has been found that:

1 faraday = 96 500 coulombs

Therefore, the quantity of electricity required to deposit one mole of copper atoms (64 g) is:

$$2 \times 96\,500 = 193\,000 \text{ coulombs (2 faradays)}$$

Examples

1 Calculate the number of faradays required to produce 10 g of silver deposited on the surface of a fork during an electroplating process (A_r: Ag = 108).
Electrode equation:

$$Ag^+(aq) + e^- \longrightarrow Ag(s)$$

1 mole of silver is deposited by 1 faraday
Therefore, 108 g of silver are deposited by 1 faraday.
Hence 1 g of silver is deposited by 1/108 faradays.
Therefore, 10 g of silver are deposited by

$$\frac{1 \times 10}{108} = 0.093 \text{ faradays}$$

2 Calculate the volume of oxygen gas, measured at room temperature and pressure, liberated at the anode in the electrolysis of acidified water by 2 faradays. (1 mole of oxygen at room temperature and pressure occupies a volume of 24 dm^3.)
Electrode equation:

$$4OH^-(aq) \longrightarrow 2H_2O(l) + O_2(g) + 4e^-$$

1 mole of oxygen gas is liberated by 4 faradays.
Therefore, 24 dm^3 of oxygen are liberated by 4 faradays.
Hence 12 dm^3 of oxygen would be liberated by 2 faradays.

3 The industrial production of aluminium uses a current of 25 000 amps. Calculate the time required to produce 10 kg of aluminium from the electrolysis of molten aluminium oxide.
Electrode equation:

$$Al^{3+}(l) + 3e^- \longrightarrow Al(l)$$

1 mole of aluminium is produced by 3 faradays.
27 g of aluminium is produced by 3 faradays.
Hence 1 g of aluminium would be produced by 3/27 faradays.

Therefore, 10 000 g of aluminium would be produced by:

$$\frac{3 \times 10\,000}{27} \text{ faradays} = 1111.1 \text{ faradays}$$

So, the quantity of electricity

$$= \text{number of faradays} \times 96\,500$$
$$= 1111.1 \times 96\,500$$
$$= 1.07 \times 10^8 \text{ coulombs}$$

$$\text{Time (seconds)} = \frac{\text{coulombs}}{\text{amps}}$$

$$= \frac{1.07 \times 10^8}{25\,000}$$

$$= 4289 \text{ seconds}$$

(71.5 minutes)

Electrolysis guidelines

The following points may help you to work out the products of electrolysis in unfamiliar situations. They will also help you to remember what happens at each electrode.

- Non-metals are produced at the anode whereas metals and hydrogen gas are produced at the cathode.
- At the anode, chlorine, bromine and iodine (the halogens) are produced in preference to oxygen gas.
- At the cathode, hydrogen is produced in preference to metals, unless unreactive metals such as copper and nickel are present.

Quick Questions

1 Write equations which represent the discharge at the cathode of the following ions:
a) K^+;
b) Pb^{2+};
c) Al^{3+};
d) H^+;
and at the anode of:
e) Br^-;
f) O^{2-};
g) F^-.
h) S^{2-}.

2 How many faradays are required to discharge one mole of the following ions:
a) Mg^{2+}
b) K^+
c) Al^{3+}
d) O^{2-}?

3 How many coulombs are required to discharge one mole of the following ions: (1 faraday = 96,500 coulombs)
a) Cu^{2+};
b) Na^+;
c) Pb^{2+}?

4 Calculate the number of faradays required to deposit 6.35 g of copper on a metal surface. (A_r: Cu = 63.5)

5 The industrial production of sodium uses a current of 15 000 amps. Calculate the time required to produce 20 kg of sodium from the electrolysis of molten sodium chloride. (1 faraday = 96 500 coulombs, A_r: Na = 23)

6 Calculate the number of faradays required to produce 10 g of gold deposited on the surface of some jewellery during an electroplating process. (A_r: Au = 197 – note gold in solution is present as the Au^{3+} ion.)

2.14 Rates of reaction

Chemical reactions occur at different rates. Some are slow, for example the decay of apples, others are very fast, for example the many chemical reactions which occur when a firework explodes.

Chemists have discovered that there are five main ways in which you can alter the rate at which a chemical reaction proceeds.

The factors which affect the **rate of a reaction** are:

- Surface area of the reactants.
- Concentration of the reactants.
- Temperature at which the reaction is carried out.
- Light.
- Catalyst.

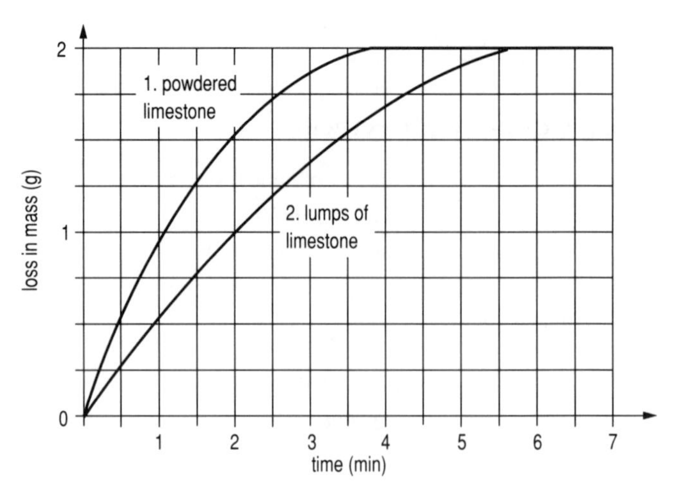

Figure 2.14.2 Sample results for the limestone/acid experiment

Surface area

Limestone (calcium carbonate) is a substance which can be used to neutralise soil acidity. Powdered limestone is used because it is found that this neutralises acidity faster than using lumps of limestone.

$$\text{hydrochloric acid} + \text{calcium carbonate} \longrightarrow \text{calcium chloride} + \text{carbon dioxide} + \text{water}$$

$$2HCl(aq) + CaCO_3(s) \longrightarrow CaCl_2(aq) + CO_2(g) + H_2O(l)$$

In the laboratory, the rate at which the reaction occurs can be followed by either measuring:

- The volume of the carbon dioxide gas which is produced.
- The loss in mass of the reaction mixture with time.

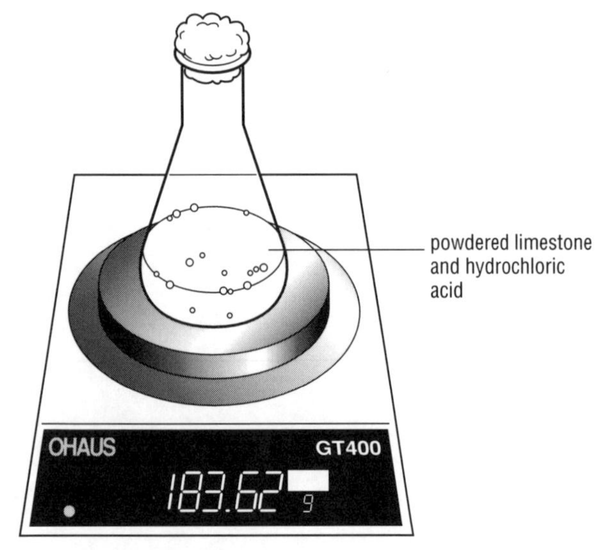

Figure 2.14.1 The total mass can be measured against time

The mass of the conical flask and the reaction mixture is measured at regular intervals. The total loss in mass is calculated for each reading of the balance, and this is plotted against time.

The reaction is generally at its fastest in the first minute. This is indicated by the slopes of the curves during this time. The steeper the slope, the faster the rate of reaction. You can see from the graphs in Figure 2.14.2 that the rate of reaction is greater with the powdered limestone.

This is because the surface area has been increased by powdering the limestone. The acid particles now have an increased amount of limestone surface with which to collide. The products of a reaction are formed when collisions occur between reactant particles. Therefore, the increase in surface area of the limestone has the effect of increasing the rate of reaction.

Concentration

In the reaction between sodium thiosulphate and hydrochloric acid, a yellow precipitate is produced.

$$\text{sodium thiosulphate} + \text{hydrochloric acid} \longrightarrow \text{sodium chloride} + \text{sulphur} + \text{sulphur dioxide} + \text{water}$$

$$Na_2S_2O_3(aq) + 2HCl(aq) \longrightarrow 2NaCl(aq) + S(s) + SO_2(g) + H_2O(l)$$

The rate of this reaction can be followed by recording the time taken for a given amount of sulphur to be precipitated. This can be done by placing a conical flask containing the reaction mixture onto a cross on a piece of paper (see Figure 2.14.3).

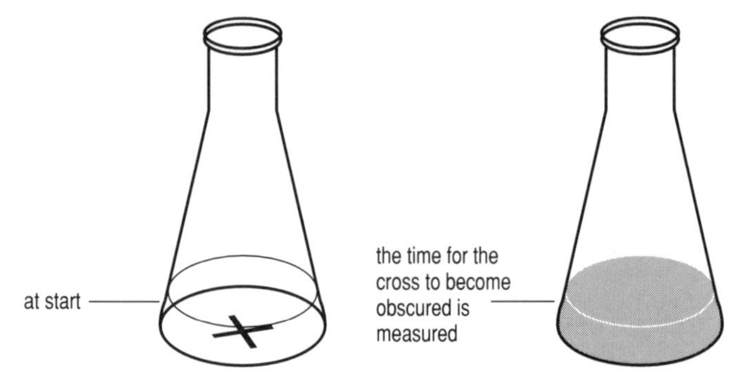

Figure 2.14.3 The precipitate of sulphur obscures the cross

As the precipitate of sulphur forms, the cross is obscured, and finally disappears from view. The time taken for this to occur is a measure of the rate of this reaction. To obtain sufficient information about the effect of changing the concentration of the reactants, several experiments of this type must be carried out, using different concentrations of sodium thiosulphate or hydrochloric acid.

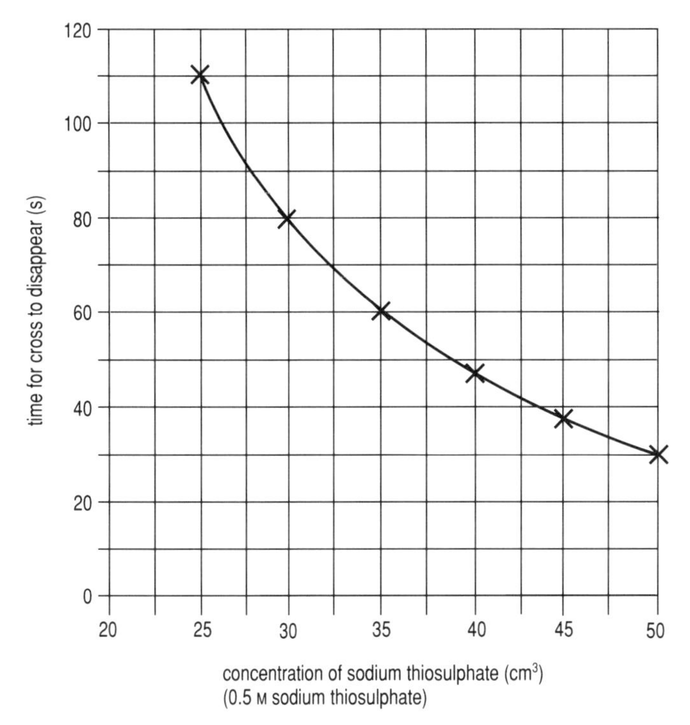

Figure 2.14.4 Sample data for the sodium thiosulphate/acid experiment at different concentrations of sodium thiosulphate

You will note from the graph that when the most concentrated sodium thiosulphate solution was used, the reaction was at its fastest. This is shown by the shortest time taken for the cross to be obscured.

The products of the reaction are formed as a result of the collisions between reactant particles. In a more concentrated solution, there are more particles and collisions occur more frequently. This means that the rate of a chemical reaction will increase if the concentration of reactants is increased.

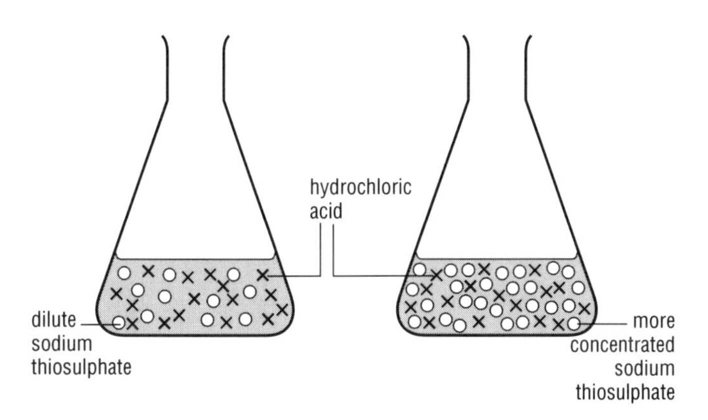

Figure 2.14.5

For reactions which involve only gases, for example the Haber process, an increase of the overall pressure at which the reaction is carried out increases the rate of the reaction. The increase in pressure results in the gas particles being pushed closer together. This means that they collide more often and so react faster.

Temperature

Food is stored in a refrigerator. This is because the rate of decay is slower at lower temperatures. This is a general feature of the majority of chemical processes. The reaction between sodium thiosulphate and hydrochloric acid can also be used to study the effect of temperature on the rate of a reaction.

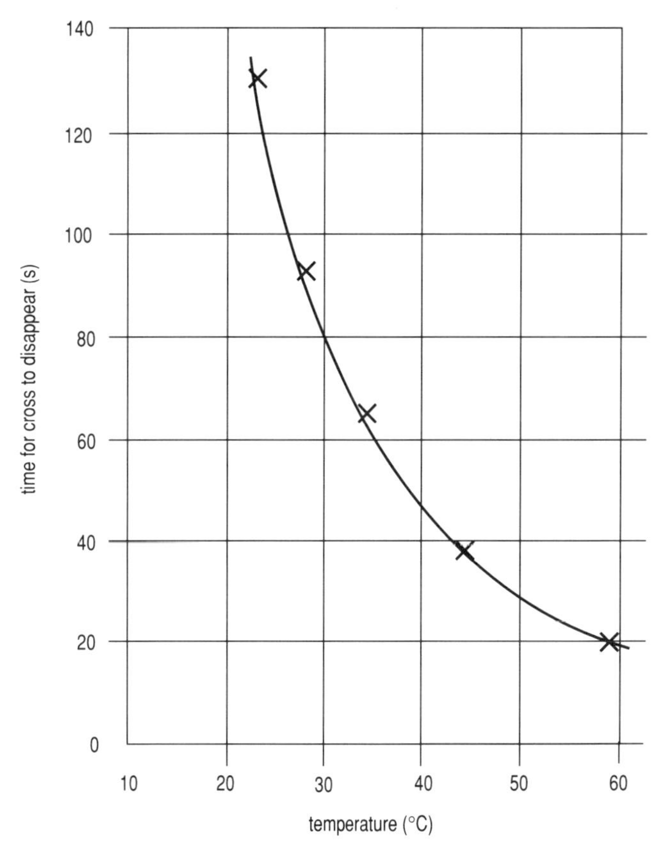

Figure 2.14.6 Sample data for the sodium thiosulphate/acid experiment at different temperatures

You can see from the graph that the rate of the reaction is fastest at high temperature. When the temperature is increased, the energy that the particles have also increases. At higher temperatures the particles move faster, increasing the number of collisions between sodium thiosulphate and hydrochloric acid. Also, because the particles have more energy, the collisions are more likely to form products. If the temperature at which a reaction takes place is increased, then the rate of reaction will usually increase.

Light

Some chemical reactions are affected by light.

Photosynthesis is a very important reaction which occurs only when sunlight falls on leaves containing the green pigment, chlorophyll.

Another chemical reaction which takes place only when exposed to light is that which occurs when you take a photograph. In this case, the silver salts on the film form silver when light falls on them.

$$\text{Silver ions} + \text{electrons} \xrightarrow{\text{light}} \text{silver}$$
$$Ag^+ + e^- \longrightarrow Ag$$

Catalysts

Over 90% of industrial processes use **catalysts**. A catalyst is a substance which can alter the rate of a reaction without being chemically changed at the end of it. The effect of a catalyst can be observed using the decomposition of hydrogen peroxide as an example. This is also a convenient way of making oxygen gas, as well as showing catalysis.

$$\text{hydrogen peroxide} \longrightarrow \text{water} + \text{oxygen}$$
$$2H_2O_2(aq) \longrightarrow 2H_2O(l) + O_2(g)$$

The rate of decomposition at room temperature is very slow. However, substances such as manganese(IV) oxide will speed up this reaction. When black manganese(IV) oxide powder is added to hydrogen peroxide solution, oxygen is produced rapidly. The rate at which this occurs can be followed by measuring the volume of oxygen gas produced with time.

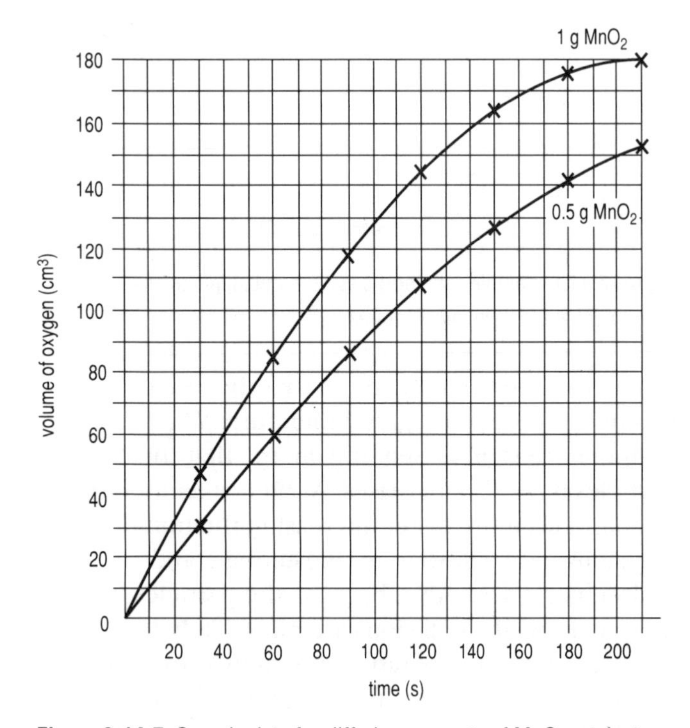

Figure 2.14.7 Sample data for differing amounts of MnO_2 catalyst

At the end of the reaction, the manganese(IV) oxide can be filtered off and used again. The reaction can be made even faster by increasing the amount and surface area of the catalyst used. This is because the activity of a catalyst involves its surface. It should be noted that, in gaseous reactions, if dirt or impurities are present on the surface of the catalyst, then it will not act as efficiently. It is said to have been 'poisoned'. Therefore the gaseous reactants must be very pure.

Chemists have found that:

- A small amount of catalyst will produce a large amount of chemical change.
- Catalysts remain unchanged chemically after a reaction has taken place, but can change physically. For example, a finer manganese(IV) oxide powder is left behind after the decomposition of hydrogen peroxide.
- Catalysts are very specific to a particular chemical reaction.

Table 2.14.1 Examples of catalysts

Process	Catalyst
Haber process – for the manufacture of ammonia	Iron
Contact process – for the manufacture of sulphuric acid	Vanadium(V) oxide
Oxidation of ammonia to give nitric acid	Platinum
Fermentation of sugars to produce alcohol	Enzymes (in yeast)
Hydrogenation of unsaturated oils to form fats in the manufacture of margarines	Nickel

A catalyst increases the rate by providing an alternative reaction path with a lower **activation energy**. The activation energy is the energy barrier which reactants must overcome, when their particles collide, to successfully react and form products (see Figure 2.14.8).

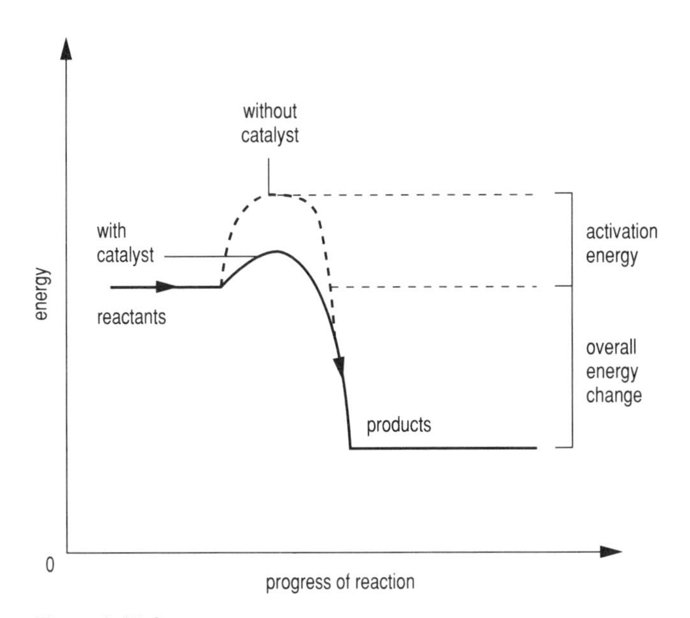

Figure 2.14.8

Enzymes

Enzymes are protein molecules produced in living cells. These substances are used by living organisms as catalysts to speed up some hundreds of different chemical reactions going on inside them. These biological catalysts are very specific, in that each chemical reaction which takes place within the living organism has a different enzyme catalysing it. You can imagine, therefore, that there are literally hundreds of different kinds of enzyme. For example, hydrogen peroxide is a substance naturally produced within our bodies (natural metabolic product). However, it is extremely damaging and must be decomposed very rapidly. Catalase is the enzyme which does this, converting hydrogen peroxide into harmless water and oxygen within our liver.

$$\text{hydrogen peroxide} \xrightarrow{\text{catalase}} \text{water} + \text{oxygen}$$

$$2H_2O_2(aq) \xrightarrow{\text{catalase}} 2H_2O(l) + O_2(g)$$

Whereas many chemical catalysts can work under various conditions of temperature and pressure as well as alkalinity or acidity, biological catalysts are very particular about the conditions. For example, they operate over a very narrow temperature range, and if the temperature becomes too high, they become inoperative. At temperatures above about 45 °C, they denature.

A huge multimillion pound industry has grown up around the use of enzymes. For example, biological washing powders contain enzymes which break down stains such as sweat, blood and egg, and they do this at the relatively low temperature of 40 °C! (This reduces energy costs because the washing water does not need to be heated to very high temperatures.)

There were problems associated with the early biological washing powders. Some customers who were allergic to the enzymes suffered from skin rashes. This problem has now been overcome to a certain extent by manufacturers advising that clothes washed in biological powders may need extra rinsing. Also, there are now warnings placed on many packets, indicating that the powder contains enzymes which may cause skin rashes.

Quick Questions

1 What apparatus would you use to measure the rate of reaction of limestone with dilute hydrochloric acid by measuring the volume of carbon dioxide produced?

2 The following results were obtained from an experiment of the type you were asked to design in question 1.

Time /mins	0	0.5	1.0	1.5	2.0	2.5	3.0	3.5	4.0	4.5	5.0
Total vol of CO_2 gas/cm^3	0	15	24	28	31	33	35	35	35	35	35

a) Plot a graph of the total volume of CO_2 against time.

b) At which point is the rate of reaction fastest?

c) What volume of CO_2 was produced after 1 min 15 secs?

d) How long did it take to produce 30 cm^3 of CO_2?

3 Devise an experiment to show the effect of changing the concentration of dilute acid on the rate of reaction between magnesium and hydrochloric acid.

4 Explain why food cooks faster in a pressure cooker than in a standard saucepan.

5 Devise an experiment to show how sunlight affects the rate of formation of silver from the silver salts, silver chloride and silver bromide.

2.15 Chemical energy

A **fuel** is a substance which can be conveniently used as a source of energy. Fossil fuels produce energy when they undergo combustion.

fossil fuel + oxygen ⟶ water + carbon dioxide + energy

For example, natural gas burns readily in air. Natural gas, like crude oil, is a mixture of hydrocarbons such as methane, ethane and propane, and may also contain some sulphur.

The perfect fuel would be:

- Cheap.
- Available in large quantities.
- Safe to store and transport.
- Easy to ignite and burn, causing no pollution.
- Capable of releasing large amounts of energy.

Solid fuels are safer than volatile liquid fuels such as petrol and gaseous fuels like natural gas. We obtain most of our energy needs from the combustion of fuels and from the combustion of foods.

Combustion

When methane burns in a plentiful supply of air it produces a large amount of energy.

methane + oxygen ⟶ water + carbon dioxide + energy

$$CH_4(g) + 2O_2(g) \longrightarrow 2H_2O(l) + CO_2(g)$$

During this process, the **complete combustion** of methane, heat is given out. It is an **exothermic** reaction. If only a limited supply of air is available then the reaction is not as exothermic and the poisonous gas carbon monoxide is produced.

methane + oxygen ⟶ carbon monoxide + water + energy

$$2CH_4(g) + 3O_2(g) \longrightarrow 2CO(g) + 4H_2O(l) + \text{energy}$$

This process is known as the **incomplete combustion** of methane.

The energy changes which take place during a chemical reaction can be shown by an energy level diagram.

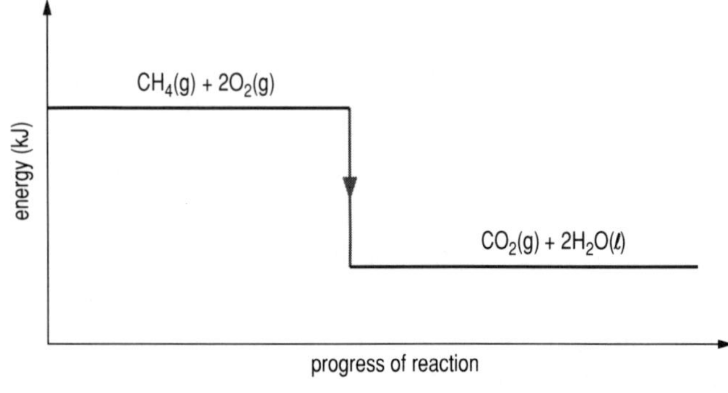

Figure 2.15.1 Energy level diagram for the complete combustion of methane

When any chemical reaction occurs, the chemical bonds in the reactants have to be broken – this requires energy. When the new bonds in the products are formed, energy is given out (see Figure 2.15.2). The **bond energy** is defined as the amount of energy, in kilojoules, associated with the breaking or making of one mole of chemical bonds in a molecular element or compound.

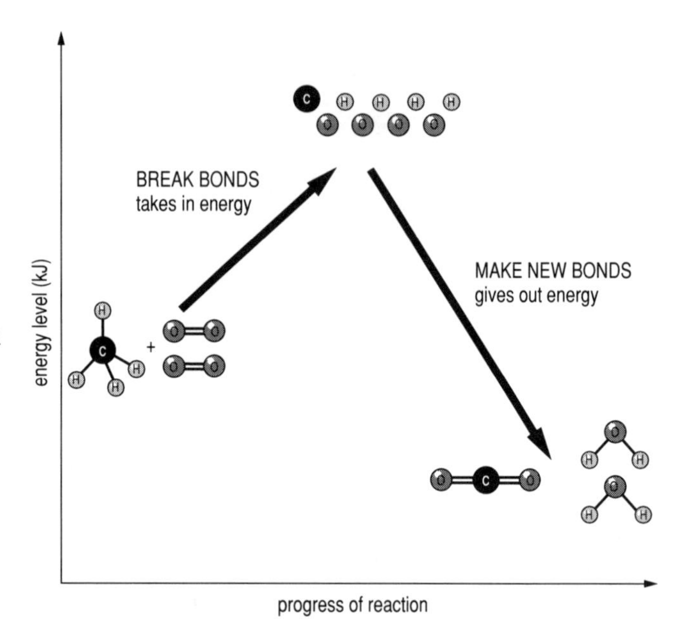

Figure 2.15.2 Breaking and forming bonds during the combustion of methane

Using bond energy data (see Table 2.15.1), which tells us how much energy is both needed to break a chemical bond and how much is given out when it forms, we can calculate how much energy is involved in each stage.

Table 2.15.1

Bond	Bond energy (kJ/mol)
C—H	435
O=O	497
C=O	803
H—O	464
C—C	347

Bond breaking

Breaking 4 C—H bonds in methane requires:
$4 \times 435 = 1740$ kJ
Breaking 2 O=O bonds in oxygen requires:
$2 \times 497 = 994$ kJ
Total 2734 kJ of energy.

Making bonds
Making 2 C=O bonds in carbon dioxide gives out:
$2 \times 803 = 1606$ kJ
Making 4 O—H bonds in water gives out:
$4 \times 464 = 1856$ kJ
Total 3462 kJ of energy.

Energy difference
= energy required to break bonds − energy given out when bonds are made
$= 2734 - 3462 = -728$ kJ

The negative sign shows that the chemicals are losing energy to the surroundings, that it is an exothermic reaction. A positive sign would indicate that the chemicals are gaining energy from the surroundings. This type of reaction would be called an **endothermic** reaction. The energy stored in the bonds is called the **enthalpy** and given the symbol ***H***. The change in energy in going from reactants to products is called the **change in enthalpy** and is shown as ΔH (pronounced 'delta H'). ΔH is called the **heat of reaction**.

For an exothermic reaction, ΔH is negative and for an endothermic reaction it is positive.

When fuels such as methane are burned, they require energy to start the chemical reaction off. This is known as the activation energy, E_A (see Figure 2.15.3). In the case of methane reacting with oxygen, it is the energy involved in the initial bond breaking (see Figure 2.15.2). The value of the activation energy will vary from fuel to fuel.

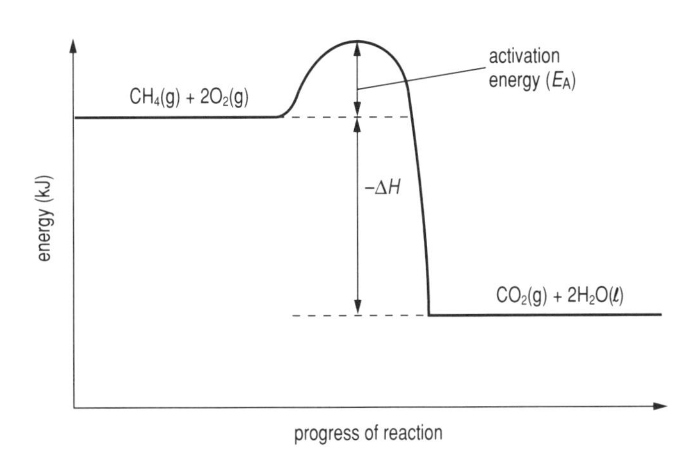

Figure 2.15.3 Energy level diagram for methane/oxygen

Endothermic reactions are much less common than exothermic ones. In this type of reaction, energy is absorbed from the surroundings so that the energy of the products is greater than that of the reactants. The reaction between nitrogen and oxygen gases is endothermic (see Figure 2.15.4).

nitrogen + oxygen ⟶ nitrogen(II) oxide

$$N_2(g) + O_2(g) \longrightarrow 2NO(g)$$

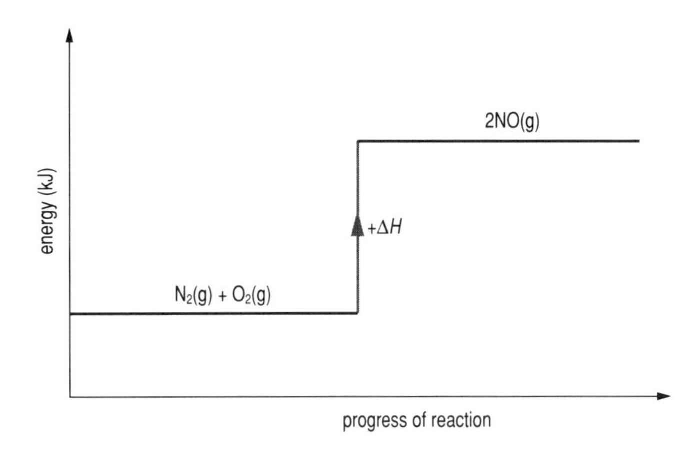

Figure 2.15.4 Energy level diagram for nitrogen/oxygen

Dissolving is often an endothermic process. For example, when ammonium nitrate dissolves in water the temperature of the water falls - indicating that energy is being taken from the surroundings.

Photosynthesis and thermal decomposition are other examples of endothermic processes.

In equations, it is usual to express the ΔH value in units of kJ/mol. For example:

$$CH_4(g) + 2O_2(g) \longrightarrow CO_2(g) + 2H_2O(l) \quad \Delta H = -728 \text{ kJ/mol}$$

This ΔH value tells us that when one mole of methane is burned in oxygen, 728 kJ of energy are released. This value is called the **enthalpy of combustion** of methane (molar heat of combustion of methane).

Enthalpy of neutralisation

This is the enthalpy change which takes place when one mole of hydrogen ions (H^+(aq)) is neutralised.

$$H^+(aq) + OH^-(aq) \longrightarrow H_2O(l) \quad \Delta H = -57 \text{ kJ/mol}$$

This process occurs, for example, in the titration of an alkali by an acid to produce a neutral solution.

Quick Questions

1 Using the bond energy data given in Table 2.15.1:
  a) Calculate the enthalpy of combustion of ethane;
  b) Draw an energy level diagram to represent this combustion process.
2 How much energy is released if:
  a) 0.5 moles of methane are burned;
  b) 4 g of methane are burned?
  (A_r: C = 12, H = 1)
3 How much energy is released if:
  a) two moles of hydrogen ions are neutralised;
  b) one mole of sulphuric acid is completely neutralised?

Section Two: Examination Questions

1. The positions of the first twenty elements in the Periodic Table are shown below.

						H											He
Li	Be											B	C	N	O	F	Ne
Na	Mg											Al	Si	P	S	Cl	Ar
K	Ca																

(a) In 1869, Dimitri Mendeléev published a table on which this Periodic Table has been based. He stated that 'when elements are arranged in order of atomic mass, similar properties recur at intervals'. Discuss whether or not this historical statement is acceptable. [3]

(b) (i) Group 1 contains the elements lithium, sodium and potassium. Why do these Group 1 elements have similar chemical properties? [1]

(ii) The order of reactivity for the Group 1 metals is:

potassium most reactive
sodium
lithium least reactive

Use their positions in the Periodic Table to explain this order of reactivity. [2]

(c) The diagram shows what happens when a small piece of potassium is placed on water. The water contains a small amount of a pH indicator.

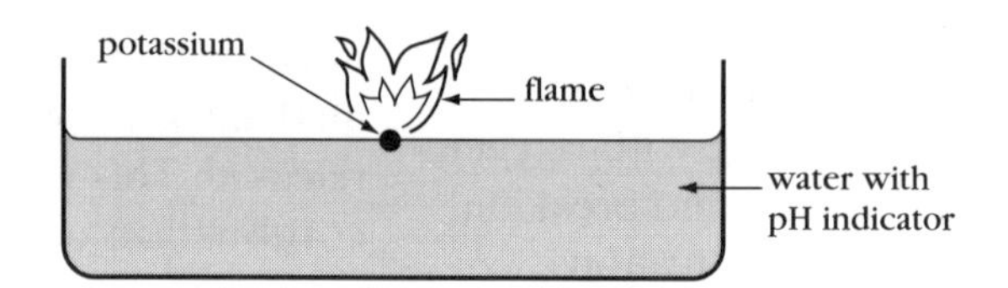

(i) What observation shows that potassium has a low density? [1]

(ii) The reaction causes the pH indicator to change colour. What has happened to the pH of the solution to cause this colour change? Explain your answer. [3]

(iii) A flame is produced when potassium reacts with the water. Explain why. [2]

(d) (i) The elements helium, neon and argon are unreactive. Explain why. [2]

(ii) In the space below draw a diagram to show the electron arrangement of an argon atom. [1]

(SEG Double Award, Higher, 1998 Specimen)

2. (a) Chlorine, bromine and iodine are three of the halogen elements that occur in Group VII of the Periodic Table. The following table gives some data for these elements.

Element	*Melting point (temperature)/°C*	*Boiling point (temperature)/°C*
Chlorine	−101	−35
Bromine	−7	59
Iodine	114	184

(i) There is another halogen element called astatine which has a higher atomic number and follows iodine in Group VII and is very unstable. Would you expect astatine to be a gas, liquid or solid at room temperature (20 °C)? [1]

(ii) Which is the **most** reactive of the three halogens shown in the table? [1]

(iii) Chlorine is a toxic (poisonous) substance, but is used on a large scale in the purification of public water supplies. Write a brief explanation of this. [2]

(b) The atoms of the elements hydrogen, chlorine and sodium, are represented by the symbols $^{1}_{1}H$; $^{35}_{17}Cl$ and $^{23}_{11}Na$, respectively.
Give the electronic structure of hydrogen, chlorine and sodium. [2]

(c) Under certain conditions, both hydrogen and sodium react vigorously with chlorine to form hydrogen chloride (HCl) and sodium chloride (NaCl) respectively.
(i) Give the electronic structure of hydrogen chloride. [2]
(ii) Show the electronic changes that take place during the formation of sodium chloride, including the structure of the product. [3]

(d) Write down with a reason, whether the structure of sodium chloride is classified as a simple molecule or a giant structure. [2]

(WJEC, Higher, 1998 Specimen)

3. This list shows some of the metals in the reactivity series:

zinc (most reactive)
iron
chromium
copper (least reactive)

In its simple chemistry, chromium exists as Cr^{3+} ions, and forms green compounds.

(a) What will you see if an excess of powdered chromium is added to aqueous copper(II) sulphate? [2]

(b) Write a balanced symbolic equation for the reaction between chromium and hydrochloric acid. [2]

(c) Chromium can be manufactured by the chemical reduction of chromium(III) oxide. Suggest a suitable reagent and the probable conditions for this reduction. [1]

(d) Electrolysis is used to coat iron car bumpers with chromium using a suitable solution containing Cr^{3+}.
(i) Predict what will happen to the iron if a chromium plated car bumper is scratched. [2]
(ii) Complete the ionic equation for the formation of chromium.
Cr^{3+} _______ $\longrightarrow$ _______ [1]

(MEG, Higher, 1998 Specimen)

4. (a) The main ore of iron is haematite which is mainly iron(III) oxide. The ore is mixed with coke and limestone and added to a blast furnace to make iron.

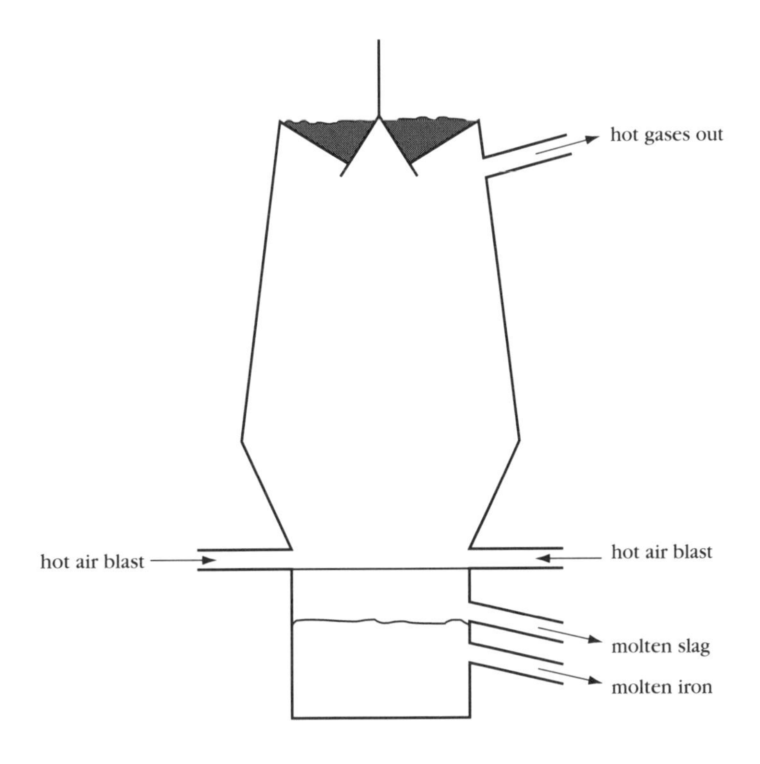

(i) What is meant by the term 'ore'? [1]
(ii) Name the **four** raw materials added to a blast furnace to produce iron. [2]

(b) The **two** reactions taking place which produce iron are:

iron(III) oxide + carbon ⟶ iron + carbon dioxide
iron(III) oxide + carbon monoxide ⟶ iron + carbon dioxide

(i) What type of reaction is the conversion of iron(III) oxide to iron? [1]
(ii) Explain the reason, for your choice of reaction type in **(b)**(i). [1]
(iii) Why is limestone added to the mixture in the blast furnace? [1]

(c) The table below shows the properties of three steels.

Steel	*% carbon*	*Properties*
Mild steel	0.25	Flexible (pliable)
Medium steel	0.40	More springy, tougher
High carbon steel	1.00	Tough but brittle

(i) Which steel would you use to make a saw blade? [1]
(ii) Explain the reason for your choice. [1]

(WJEC, Higher, 1998 Specimen)

5. A bottle of Irish Mineral Water has the following information on the label.

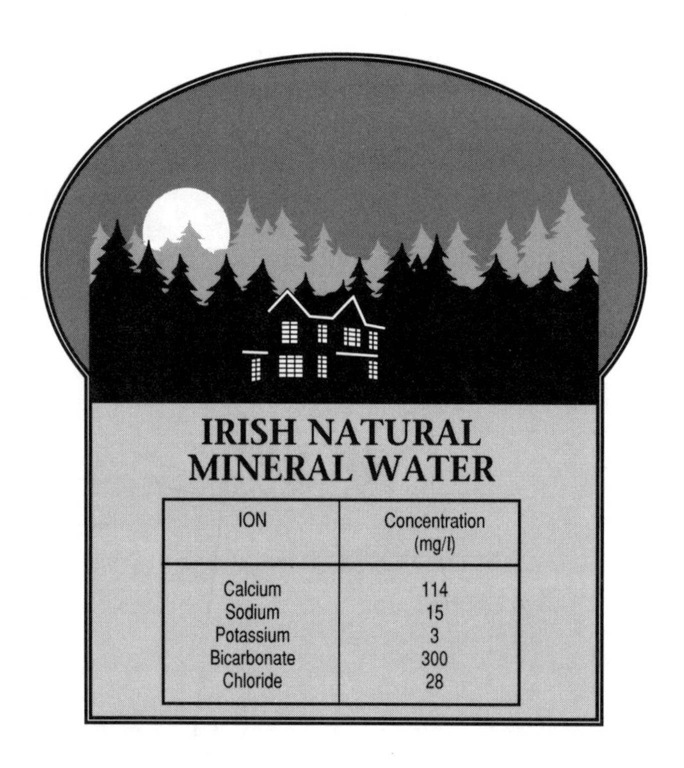

ION	Concentration (mg/l)
Calcium	114
Sodium	15
Potassium	3
Bicarbonate	300
Chloride	28

(a) How could you tell from the label that this mineral water could come from a hard water region? [1]
(b) Give **one** benefit to health of drinking hard water. [1]
(c) Name **one** physical characteristic you would expect to find in a hard water region. [1]
(d) What is the difference between temporary hardness and permanent hardness? [2]

(NICCEA Double Award, Higher, 1998 Specimen)

6. (a) An experiment to investigate the electrolysis of concentrated sodium chloride solution was set up as shown in the diagram below.

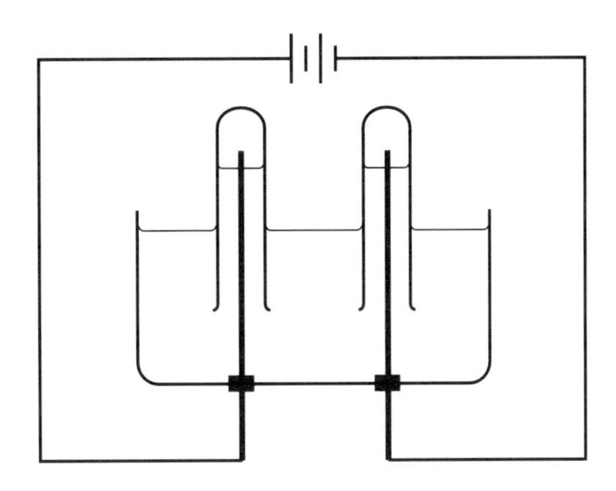

(i) What is meant by the term 'electrolysis'? [2]
(ii) What name is given to the negative electrode? [1]
(iii) What is observed at the **anode** during the above electrolysis experiment? [2]
(iv) What is observed at the negative electrode during the above electrolysis experiment? [2]
(v) Write a **balanced ionic** equation for the reaction taking place in (iv) above. [2]

(b) (i) Impure copper metal can be purified using electrolysis. Describe how this may be carried out. [5]
(ii) Write a **balanced ionic** equation to show the formation of copper in **(b)**(i) above. [2]

(NICCEA, Higher, 1998 Specimen)

7. An experiment was carried out to compare the speeds of the reactions of magnesium lumps and magnesium powder with dilute hydrochloric acid. Equal masses of lumps and powder were used. A fresh sample of the same dilute hydrochloric acid was used for each reaction.

(a) Suggest **two** other things which must be kept the same in the two reactions for the experiment to be a fair test. [2]

(b) Complete a word equation for the reaction.

_______ + _______ $\longrightarrow$ _______ + _______ [2]

(c) In each reaction the total volume of gas collected was measured at regular intervals. Draw a labelled diagram of apparatus which could be used to carry out the experiment. [2]

(d) The graph shows the results of the reaction of the **lumps** and the acid.

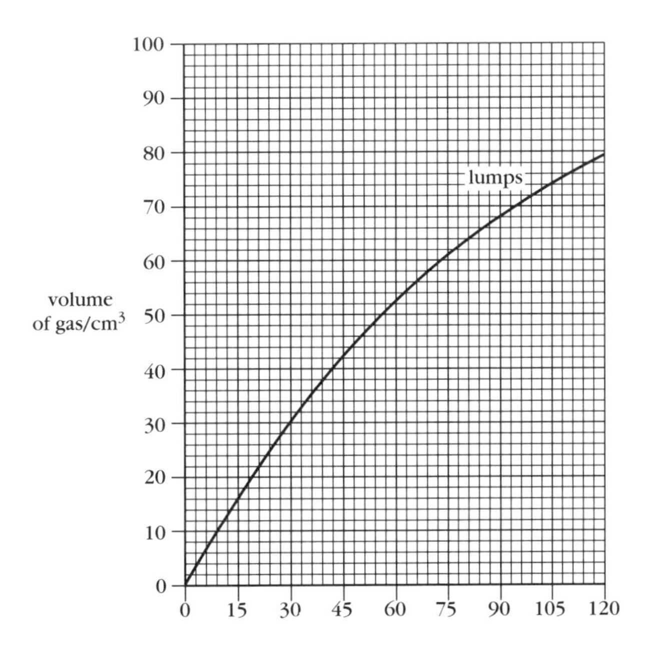

The results of the reaction between the **powder** and the acid are shown below.

time/s	0	15	30	45	60	75	90	105	120
total volume of gas/cm^3	0	50	70	80	90	96	99	100	100

(i) Plot these results on the same grid. Draw a curve to fit the points, and label it 'powder'. [2]
(ii) What conclusion can you draw from this experiment? [1]
(iii) Explain your answer to (ii) in terms of collisions of particles. [2]
(iv) The reaction between magnesium and hydrochloric acid is exothermic. This could mean that the experiment was not a fair test.
Explain why not. [2]

(MEG, Higher, 1998 Specimen)

8. A class studied five different metals, labelled A, B, C, D and E. Some of the experiments were carried out by the teacher and some by the pupils. The results of the experiments are shown in the table.

METAL	RESULT OF HEATING IN AIR	REACTION WITH COLD WATER	REACTION WITH STEAM	REACTION WITH DILUTE ACID
A	Did not burn. Oxide formed on surface	No reaction	Slow reaction	Slow reaction
B	No reaction	No reaction	No reaction	No reaction
C	It burned violently Oxide formed	Bubbles of hydrogen were quickly produced	NOT ATTEMPTED	NOT ATTEMPTED
D	Did not burn Oxide formed on surface	No reaction	No reaction	No reaction
E	It burned quickly. Oxide formed.	A few bubbles on the surface of the metal.	Vigorous reaction. Hydrogen and oxide produced.	Dissolved quickly. Hydrogen produced.

(a) Use the results of the experiments to place the five metals, A, B, C, D and E, in an order of reactivity. [4]
(b) Why were the reactions of metal C with steam or with acid **not** carried out? [1]
(c) At the end of the lesson the teacher told the class that the five metals were calcium, copper, gold, iron and magnesium.
Use information from the Chemistry Data Tables at the back of the book to identify each of the metals A, B, C, D and E. [1]

	Name of metal
A	
B	
C	
D	
E	

(d) Explain in terms of their structure, **why** metals are good conductors of heat and electricity [3]

(NEAB, Higher, 1998 Specimen)

9. (a) Aluminium is **not** found in the Earth's crust as the pure metal. It is obtained from an ore called bauxite which is impure aluminium oxide.
Explain why aluminium is not found in the Earth's crust as the pure metal. [2]

(b) A labelled diagram of a cell used to produce aluminium metal is shown below.

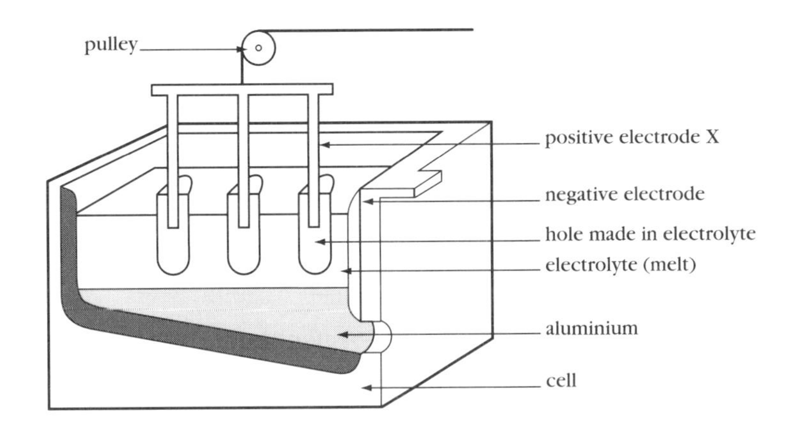

(i) What is X? [1]
(ii) Give a balanced ionic equation for the reaction which takes place at the cathode. [1]
(iii) A gas is made in the cell as well as the aluminium. What is this gas? [1]
(iv) X is hung on a pulley so it can be lowered. Suggest why X might have to be lowered while aluminium is being produced. [1]
(v) Which has the higher density, aluminium metal or the melt?
Give the reason for your answer. [1]

(c) High purity copper is obtained by electrolysis using a thin, pure copper cathode and a solution of copper sulphate. A current of 200 amperes (A) is used for 12 hours. What mass of copper is formed at the cathode?
Include in your answer the equation you are going to use. Show clearly how you obtain your answer and give the unit.
(The Faraday constant (*F*) is 96 500 coulombs per mole (C/mol). The relative atomic mass of copper is 64.) [5]

(SEG, Higher, 1998 Specimen)

10. The salt sodium hydrogen phosphate, (Na_2HPO_4), is used as a softening agent in processed cheese. The salt can be made by reacting phosphoric acid (H_3PO_4) with an alkali.

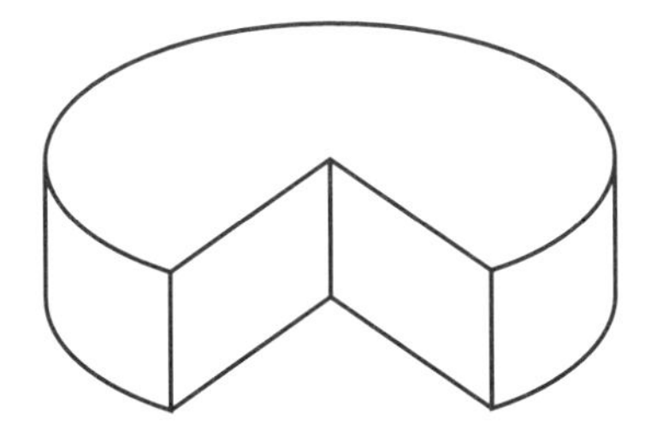

(a) Complete the name of an alkali that could react with phosphoric acid to make sodium hydrogen phosphate.
.... hydroxide [1]

(b) What name is given to a reaction in which an acid reacts with an alkali to make a salt? [1]

(c) Use the Chemistry Data Tables at the back of this book to answer these questions.
(i) What colour is universal indicator in pure water? [1]
(ii) A solution of phosphoric acid was tested with universal indicator solution. The indicator changed colour to orange.
What was the pH of the phosphoric acid solution? [1]

(d) How would the pH change when alkali is added to the phosphoric acid solution? [1]

(e) (i) What ions are present when any acid is dissolved in water? [1]

(ii) What ions are present when any alkali is dissolved in water? [1]

(iii) Write a chemical equation for the reaction that takes place between the ions named in (i) and (ii). [1]

(NEAB, Foundation)

11. Describe **chemical** tests by which you could distinguish between the following pairs of substances. Give one test for each pair and state what you SEE.

(a) 'Common salt' (sodium chloride) and 'sugar' (sucrose). [3]

(b) Dilute sulphuric acid and water. [3]

(c) Sodium carbonate and sodium sulphate. [3]

(London, Higher, 1998 Specimen)

12. Many reactions proceed by the breaking of covalent bonds followed by the formation of new bonds. Use this information, where it is helpful, to explain the following.

(a) (i) The reaction between hydrogen and oxygen, is very exothermic.

(ii) Hydrogen and oxygen do not react when left together at room temperature. [3]

(b) The decomposition of ammonia into its elements is an endothermic process. [3]

(c) The reaction represented by the equation

$$4P(g) \longrightarrow P_4(g)$$

is exothermic. [2]

(London, Higher, 1998 Specimen)

Section THREE

3.1 The structure of the Earth

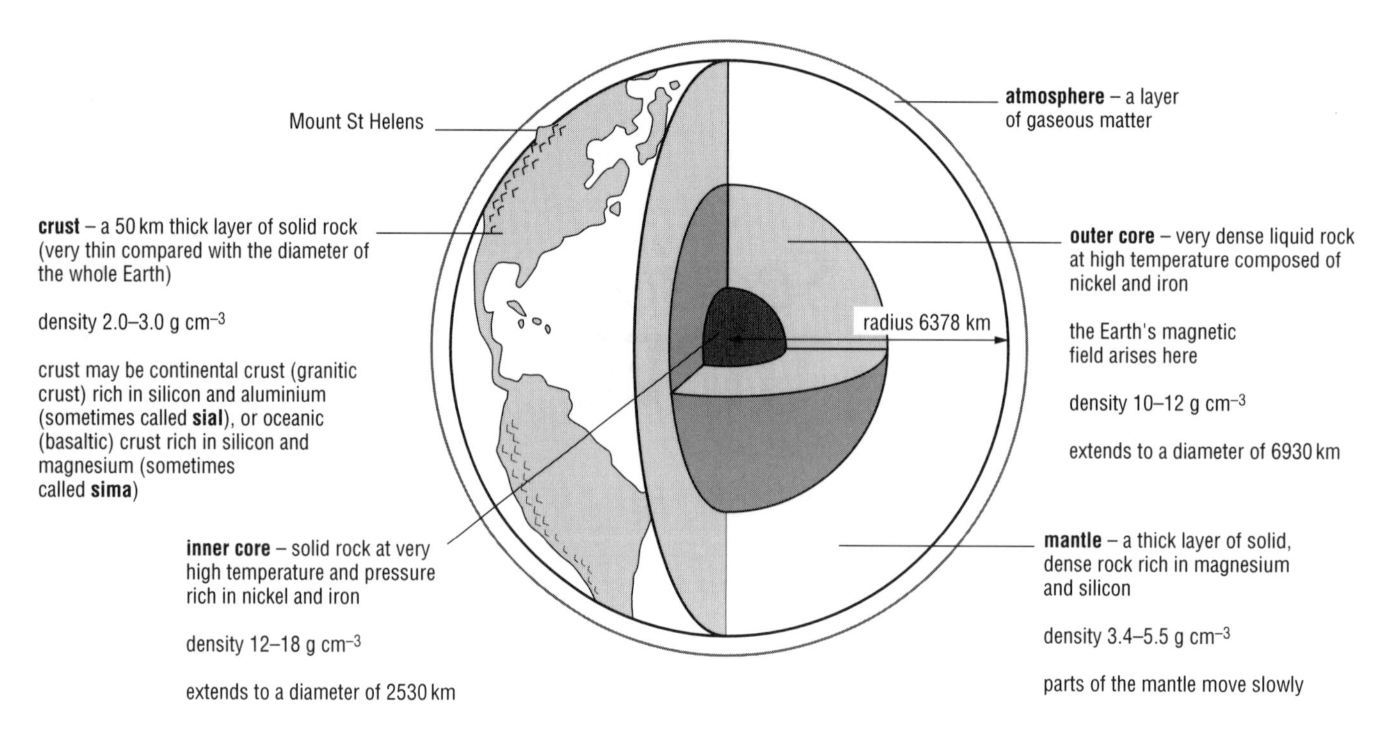

Figure 3.1.1 Planet Earth

Scientists have used sound waves as well as the waves sent out by earthquakes to study the structure of the Earth.

The **core** has a temperature of about 4300 °C. This temperature drops as you go into the **mantle** and just below the crust, the temperature is only about 1000 °C! These high temperatures are maintained mainly by:

- The inside of the Earth being insulated by the outer layers.
- The radioactive isotopes of the elements potassium, thorium and uranium. The nuclei of these isotopes are unstable and break up, giving out large amounts of energy as they change into smaller nuclei.

What is the crust made of?

There are many different rocks in the Earth's crust.

Igneous rocks

Igneous rocks are formed when hot magma from the Earth's mantle cools and hardens. Igneous rocks are usually crystalline. There are two main types of igneous rock:
Intrusive - formed by crystallisation of the magma underground. An example is granite.
Extrusive - formed by crystallisation of the magma on the Earth's crust. An example is basalt.

Sedimentary rocks

Sedimentary rocks cover approximately 75% of the continents. They are formed by **deposition** when solid particles carried or transported in seas and rivers are deposited. Layers of sediment can pile up for millions of years and the sediment at the bottom of the pile experiences great pressure, cementing the grains together. Sedimentary rocks have definite layers or strata, which can often be seen. There is a large variation in their hardness and grain size, and sedimentary rocks often contain **fossils**.

Limestone is a sedimentary rock which formed beneath the sea. It is now found well above sea level due to the movement of the Earth's crust during the process of uplift. Uplift occurs mainly because of the large scale lateral forces at work on the Earth's crust, resulting in its crumpling, for example, at plate boundaries.

Limestone is composed mainly of calcium carbonate which effervesces when it comes into contact with a dilute acid. This property is often used to show the presence of limestone in a rock sample. Sandstone is another example of a sedimentary rock.

Metamorphic rocks

Metamorphic rocks are formed when rocks buried deep beneath the Earth's surface are altered by the action of great heat and pressure. Marble is a metamorphic rock and is formed by this type of action on limestone. Slate is another example of a metamorphic rock which is formed from mudstone.

Fossils

Fossils are the remains or impressions of living organisms, made when animals or plants die in soft **silt**. When these organisms decay they leave their impressions in the silt as it is turned into rock through the action of temperature and pressure. In some cases the organism decays and dissolves in the silt, leaving a space. When this happens, certain minerals may seep into the space and take up the shape of the dissolved organism. When the silt turns into rock, a fossil 'mould' is preserved.

Fossils are found in sedimentary rocks such as limestone. The older sedimentary rocks are found deeper underground than the younger rock. Geologists, by examining the fossils found in the different rock strata, have been able to divide time into three **eras** according to the type of fossil found.

Cainozoic era
This is the most recent era and covers the present, to 65 million years ago.

Mesozoic era
This era covers the time from 65 million years ago to 225 million years ago.

Palaeozoic era
This era covers the time from 225 million years ago to 570 million years ago.

The rock cycle

The rock types discussed above interchange over millions of years in processes described by the **rock cycle**.

This concept was first developed by James Hutton. In this cycle, rocks on upland areas are weathered (see Section 3.3) and the particles are carried away by erosion to form sediments which eventually become sedimentary rocks. These are then brought to the surface by the Earth's movements (uplifting), or they may be heated and compressed to form metamorphic rocks. If these metamorphic rocks are pushed deep below the surface they will melt, forming magma in the mantle. The magma may then be squeezed upwards to the surface and igneous rock are formed. So one type of rock may be recycled to form another type of rock.

Quick Questions

1. Describe the different processes involved in the rock cycle by constructing a flow diagram of the changes which may occur.
2. What is a fossil, and how might their discovery enable geologists to date rock?
3. What is:
   a) Igneous rock;
   b) Sedimentary rock;
   c) Metamorphic rock?
   Give an example of each type of rock.

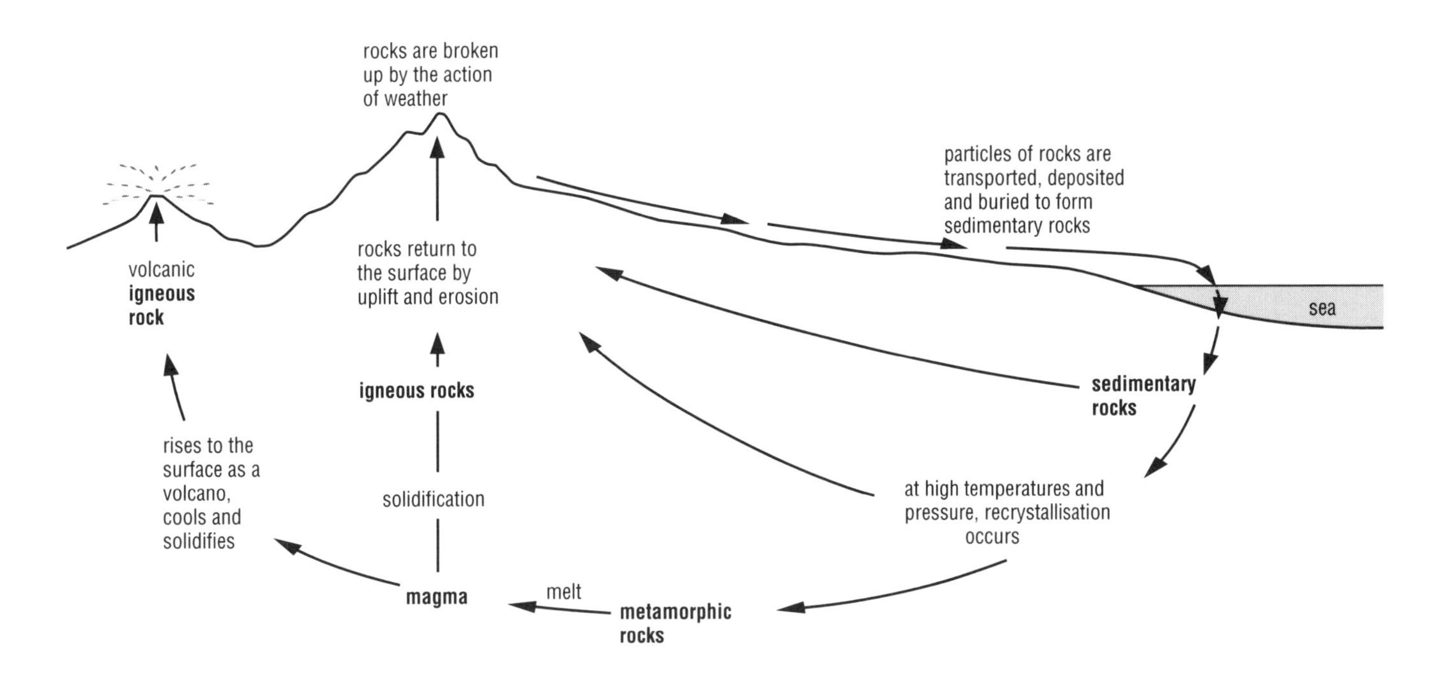

Figure 3.1.2 The rock cycle

3.2 The moving Earth

Rocks in the Earth's crust are less dense than those below them, so they float on the partially molten mantle underneath. There are two types of crust – the crust under the continents is called **continental crust**, and the crust under the oceans is known as **oceanic crust**. The continental crust contains lighter rocks such as granite, whereas the oceanic crust contains denser rocks such as basalt.

Evidence from geologists shows that the Earth's crust is not a continuous structure but in fact is divided into sections called tectonic **plates** (see Figure 3.2.1). These plates are actually moving very slowly about the Earth's surface. The driving force behind this movement is thought to be **convection currents** within the mantle.

The study of the way these plates behave is called **plate tectonics**. Geologists thought that the continents were originally joined together into one giant continent which they called Pangea.

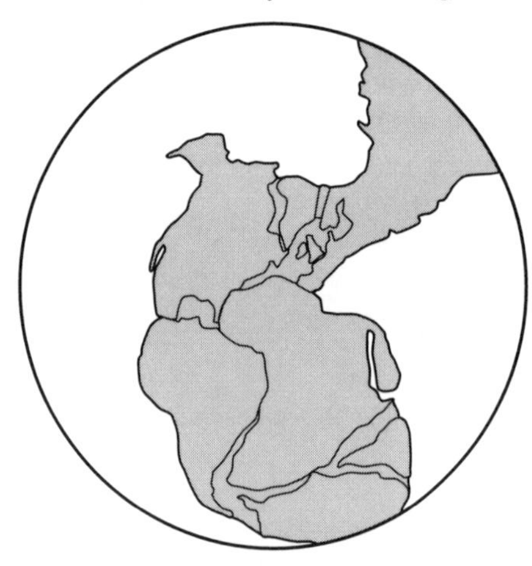

Figure 3.2.1 The giant continent was called Pangea

It is thought that it has taken approximately 200 000 000 years for the continents, as we know them today, to drift to the positions they are now in. Evidence for this has been obtained from fossils found in the U.S.A. These fossils show that animals and plants found in the U.S.A. 200 000 million years ago also lived in Europe about that time.

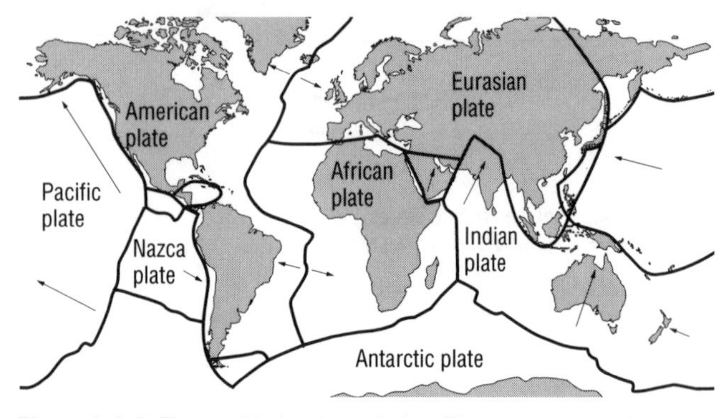

Figure 3.2.2 The world plate boundaries. The arrows show the directions in which the plates are moving

Where plates join, enormous forces are generated. This can, and does, create **earthquakes** as well as volcanoes and mountains.

If you look closely at the diagram above, you will recognise that where the plates meet are the regions where volcanic and earthquake activity occurs. Earthquakes may be caused in one of two ways:

- By the plates scraping past each other on a 'fault'.
- At margins where one plate is descending into the mantle.

Geologists have monitored earthquake activity around the Earth through hundreds of seismic stations. These stations are able to detect earthquake waves using an apparatus called a **seismometer**.

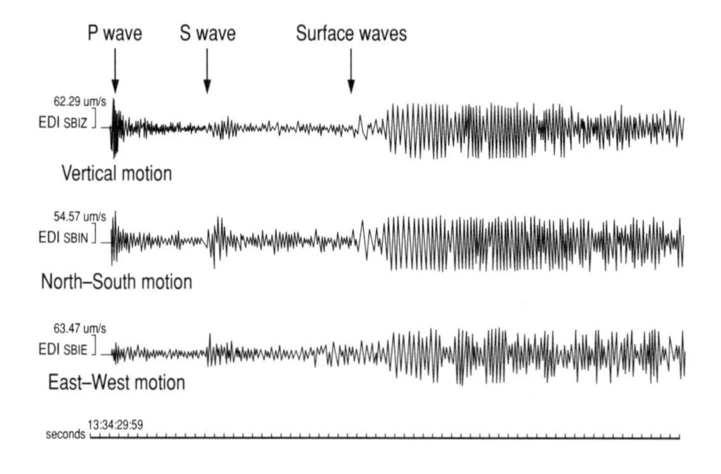

Figure 3.2.3 An example of a seismogram

There are three types of waves produced by earthquakes:

Primary (P) waves

These are longitudinal waves, produced by pushing and pulling forces which cause the rock to shake backwards and forwards. This type of wave travels at speeds of up to 6 km per second and travels through both solids and liquids.

Secondary (S) waves

These are transverse waves and cause the rock to shake at right angles to the direction of its movement. This type of wave travels at speeds of up to 3 km per second and travels only through solids.

Surface waves

This type of wave has a long wavelength and is responsible for the largest land movements and, therefore, causes the most damage. This type of wave travels more slowly than either the P or S wave.

The actual power of an earthquake is measured using the Richter scale. Each unit on this scale represents a ten fold increase on the previous unit. In theory, the scale does not have a limit. However, earthquakes are rarely encountered above 8.

Richter scale

Unit	Destruction level
2-3	Hardly noticed
3-4	Slightly noticed
4-5	Minor
5-6	Damaging
6-7	Destructive
7-8	Major destruction
8+	Enormously destructive

The meeting of the plates also leads to the formation of high mountain ranges, for example the Himalayas. This is where the crust is at its weakest and most unstable.

The spreading oceans

Although it may be difficult to believe, there are mountain ranges under the sea which are much higher than those on land. They occur when two plates are moving apart. One such region lies in the middle of the Atlantic Ocean (see Figure 3.2.4).

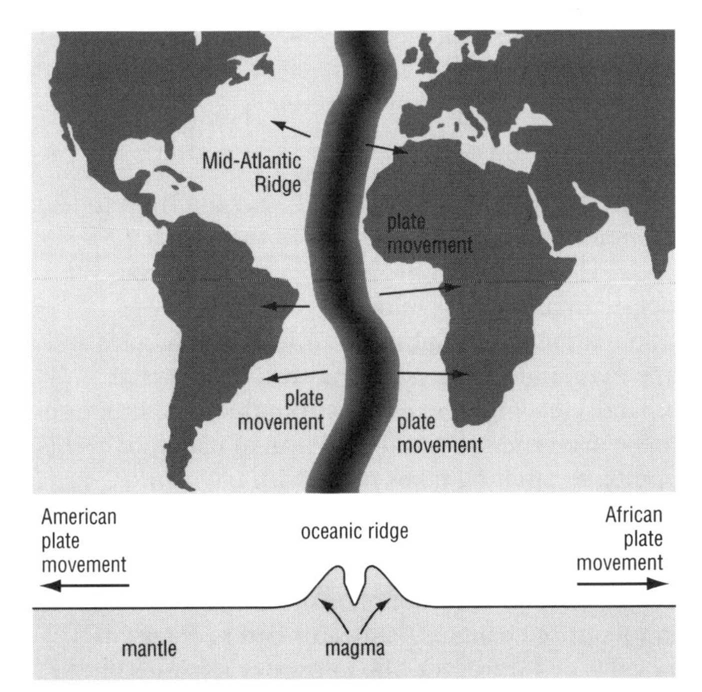

Figure 3.2.4 Formation of the Mid-Atlantic Ridge

Along the line where the plates meet, lava pours through huge fissures, or cracks. The magma cools quickly in the colder depths of the ocean, forming 'pillows'. These are quickly burst by the pressure of the lava below, and fresh lava erupts on top of the previously erupted and now hardened layer. Layer upon layer of basalt is formed, creating a huge ridge called the Mid-Atlantic Ridge. As more magma emerges, the older rock layers are pushed further and further apart - a process called **ocean-floor spreading**.

Magnetic stripes

On each side of ocean ridges, the rocks show a clear pattern of magnetic stripes. These are caused by iron crystals in the magma which, as it hardens, line up in the direction of the magnetic North and South poles. The iron particles themselves are weakly magnetised, with their North and South poles aligned to the Earth's magnetic field.

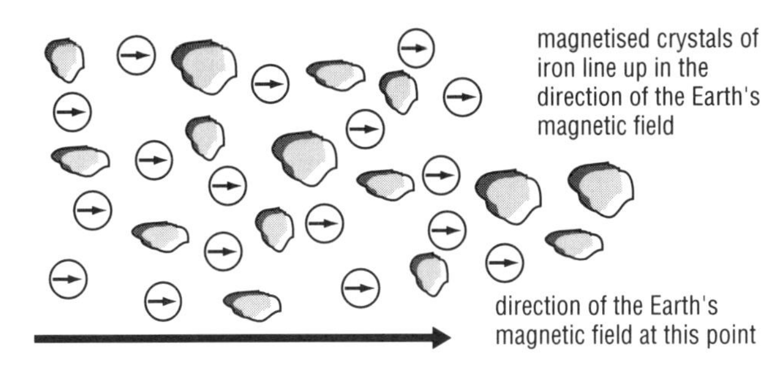

Figure 3.2.5 The iron crystals line up with the Earth's magnetic field

At times during the Earth's history, the magnetic North and South poles have suddenly reversed. The reversal of the magnetically aligned particles in the stripes confirms this, but also shows that the basalts on each side of the ridge were intruded into the ridge and became magnetised before being broken in two and moving apart, see Figure 3.2.6.

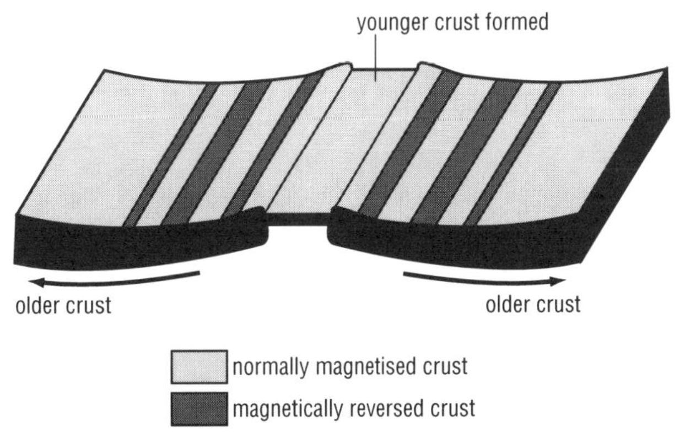

Figure 3.2.6 Magnetic stripes in the Earth's crust

Quick Questions

1. Fossils of sea creatures are found high up on Mount Everest. What does this suggest about how the Himalayan mountain range was formed? Explain your answer.
2. What evidence could be gathered to show that the east coast of America, and the west coast of Europe, were once joined?

3.3 The changing Earth

Volcanoes

In the crust, rocks are pushed down by the weight of the rocks above them when tectonic plates move towards each other. As the rocks are pushed down into the Earth they become molten due to the high temperatures which exist at these lower levels of the crust. These molten rocks are known as **magma**. Because the magma is a liquid it is less dense than the surrounding rock and so begins to rise upwards towards the surface of the crust. This molten rock material, containing dissolved gases and water beneath the Earth's crust, escapes to the surface through any areas of weakness in the crust. These weaknesses can be in the form of **fissures** or **vents** (holes). The magma appears at the surface as **lava** (see Figure 3.3.1).

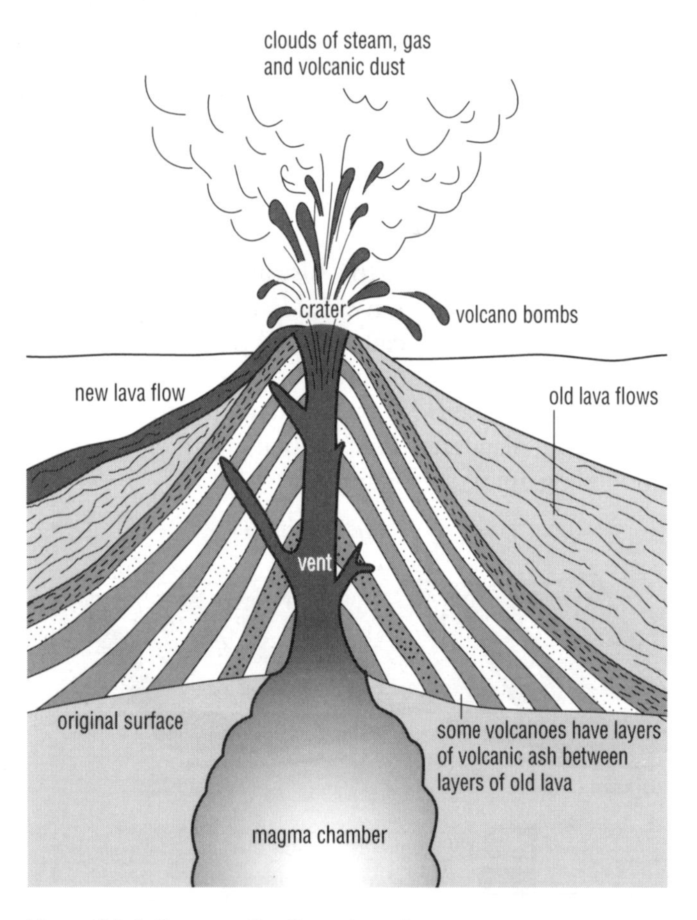

Figure 3.3.1 Cross-section through a volcano

Lava flow can engulf vast areas of land around the volcano. Some volcanoes also produce toxic gases such as sulphur dioxide and hydrogen sulphide.

Extrusive igneous rocks are formed when magma bursts through the Earth's crust. They cool rapidly, forming rocks such as basalts. Because they cool rapidly, the crystals in these rocks are so small they often cannot be seen. The rock may look smooth and glossy.

Figure 3.3.2 Different sized crystals in extrusive (basalt, *left*) and intrusive (granite, *right*) igneous rock

At other times, molten rock may be forced up into the Earth's crust but be unable to break through to the surface. These are **intrusive igneous rocks**. Because these rocks cool more slowly, the crystals formed are much larger and can be seen. This is the case in granite.

Weathering

Weathering is the breakdown of exposed rock on the Earth's surface. There are two main ways this happens - chemically or physically.

Chemical means

Rain water contains dissolved carbon dioxide as well as other gases such as sulphur dioxide and nitrogen dioxide. This **acid rain** can dissolve rocks such as limestone quite easily.

$$\text{calcium carbonate} + \text{carbon dioxide} + \text{water} \longrightarrow \text{calcium hydrogencarbonate}$$

$$CaCO_3(s) + CO_2(aq) + H_2O(l) \longrightarrow Ca(HCO_3)_2(aq)$$

Minerals may also be oxidised. Oxygen from the air can combine with iron silicates to form iron(III) oxide. This leads to a brown stain on the surface of rocks containing this mineral.

Some minerals combine with water molecules and take them into the crystal structure. The crystal expands, leading to stresses within the rock structure. This causes the rocks to break up. An example of this type of weathering takes place with haematite (Fe_2O_3). With water it forms limonite ($Fe_2O_3.H_2O$).

Physical means

This is caused by the action of wind, water and temperature changes. Rainwater enters cracks in rocks. When it freezes, the rainwater expands and this increase in volume forces the rock apart.

Stresses can also be built up by temperature changes. Minerals within the rock expand and contract with temperature changes at different rates. In temperate areas of the world such as Britain, where alternate freezing and thawing happens frequently, pieces of rock break off and fall down hill and mountainsides forming scree.

Erosion

Erosion involves the wearing away of rock, and its transportation to another place. The diagrams below show the four main ways by which erosion takes place.

Figure 3.3.3a This deep gorge was formed by the eroding action of the river

Figure 3.3.3b This stack was formed by wave action

Figure 3.3.3c This **glacier** is eroding the mountain it is attached to

Figure 3.3.3d Wind erosion caused these formations

Taking into account the four main methods of erosion, it has been found that the erosion rate for a land area lies between 8 and 9 cm of depth per 1000 years.

Quick Questions

1. What is the difference between weathering and erosion?
2. Explain the meaning of the following terms:
   a) Magma;
   b) Fissure;
   c) Lava.

3.4 Limestone

Limestone is one form of calcium carbonate ($CaCO_3$), but it is also found naturally as chalk, calcite and marble.

Deposits of chalk were formed from the shells of dead marine creatures that lived many millions of years ago. In several places in Great Britain, the chalk was covered with other types of rock and was, therefore, put under great pressure. This caused the relatively soft chalk to be changed into the harder material, limestone.

Marble, one of the other major forms of calcium carbonate, was formed in places where the chalk was not only subjected to a high pressure but also a high temperature.

Although limestone is cheap to **quarry** as it is found near the Earth's surface, there are some environmental costs in extracting it.

Uses of limestone

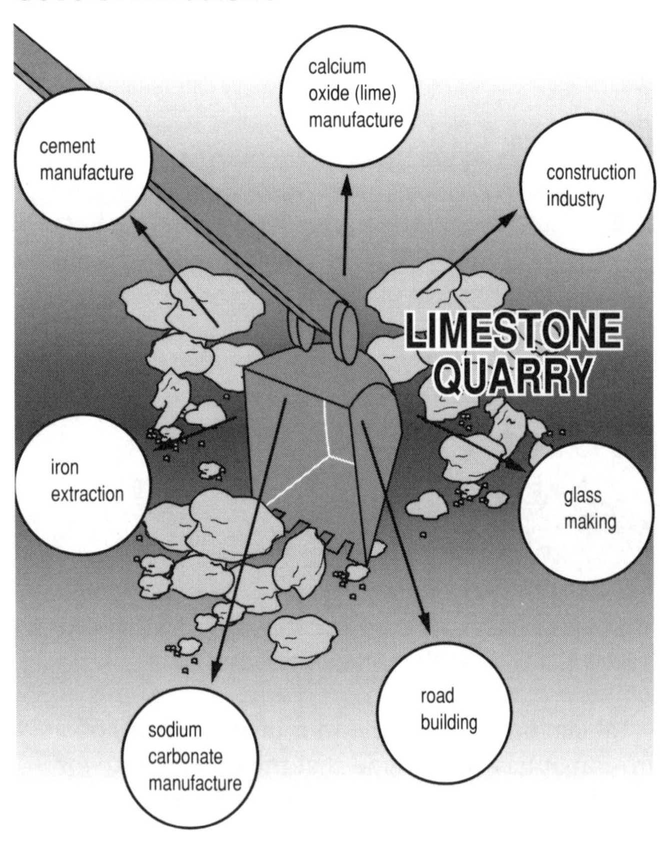

Figure 3.4.1

Direct uses of limestone

Neutralising acid soil

The reaction of limestone with acidic soil can be shown by the ionic equation:

carbonate ion (from $CaCO_3$)	+	hydrogen ion (from acid soils)	⟶	carbon dioxide	+	water
$CO_3^{2-}(s)$	+	$2H^+(aq)$	⟶	$CO_2(g)$	+	$H_2O(l)$

Manufacture of iron and steel

In the blast furnace, limestone is used to remove earthy and sandy materials found in the iron ore to form a liquid slag (see Section 2.5).

Manufacture of cement and concrete

Limestone (or chalk) is mixed with clay (or shale) in a heated rotary kiln, using coal or oil as the fuel. The material produced is called cement. It contains a mixture of calcium aluminate ($Ca(AlO_2)_2$) and calcium silicate ($CaSiO_3$). The dry product is ground to a powder and then a little calcium sulphate ($CaSO_4$) is added to slow down the setting rate of the cement. When water is added to the mixture, complex chemical changes occur and a hard, interlocking mass of crystals of hydrated calcium aluminate and silicate is formed.

Concrete is a mixture of cement with stone chippings and gravel, which gives it body. Reinforced concrete is made by setting concrete around steel rods or mesh to give the greater **tensile strength** required for the construction of large bridges and tall buildings.

Manufacture of sodium carbonate – the Solvay process

Sodium carbonate (Na_2CO_3) is one of the world's most important industrial chemicals. It is used in the manufacture of soaps, detergents, dyes, drugs and other chemicals. Sodium carbonate is manufactured ingeniously using calcium carbonate, sodium chloride, carbon dioxide and ammonia in a continuous process. Both the carbon dioxide and ammonia are recycled continuously.

Indirect uses of limestone

Lime manufacture

When calcium carbonate is heated strongly, it **thermally dissociates** (breaks up reversibly) to form calcium oxide and carbon dioxide.

calcium carbonate	⇌	calcium oxide	+	carbon dioxide
$CaCO_3(s)$	⇌	$CaO(s)$	+	$CO_2(g)$

This reaction can go in either direction, depending on the temperature and pressure used. It is an important industrial process and takes place in a lime kiln. The calcium oxide produced from this process is known as 'quicklime' or **'lime'**.

Calcium oxide is a base and some farmers still spread it on fields to neutralise soil acidity. It also has a use as a drying agent in industry. Soda glass is made by heating sand with soda (sodium carbonate, Na_2CO_3) and lime.

Large amounts of calcium oxide are also converted into calcium hydroxide ($Ca(OH)_2$) which is called **slaked lime**.

Manufacture of calcium hydroxide – slaked lime

Calcium hydroxide is a cheap industrial alkali. It is used in large quantities to make bleaching powder, by some farmers to reduce soil acidity, in the manufacture of whitewash, glass manufacture and in water purification. Calcium hydroxide, a white powder, is produced by adding an equal amount of water to calcium oxide.

This process can be shown on a small scale in the laboratory by heating a lump of limestone very strongly to convert it to calcium oxide. Water can then be carefully added dropwise, to the calcium oxide. An exothermic reaction takes place as the water and calcium oxide react together (**slaking**) to form calcium hydroxide.

calcium oxide + water ⟶ calcium hydroxide

$$CaO(s) + H_2O(l) \longrightarrow Ca(OH)_2(s)$$

A weak solution of calcium hydroxide in water is called limewater. It is used to test for carbon dioxide gas as a white solid of calcium carbonate is formed if carbon dioxide gas is mixed with it.

calcium hydroxide + carbon dioxide ⟶ calcium carbonate + water

$$Ca(OH)_2(aq) + CO_2(g) \longrightarrow CaCO_3(s) + H_2O(l)$$

Calcium hydroxide is mixed with sand to give mortar. When mixed with water and allowed to set, a strongly bonded material is formed which is used to hold bricks together. The hardening of mortar takes place as the following reaction occurs.

calcium hydroxide + carbon dioxide ⟶ calcium carbonate + water

$$Ca(OH)_2(aq) + CO_2(g) \longrightarrow CaCO_3(s) + H_2O(l)$$

Carbonates and hydrogencarbonates

Carbonates are an important range of compounds. They are all salts of carbonic acid (H_2CO_3). Many of these occur naturally in rock formations. For example, in addition to limestone, chalk and marble, malachite is copper(II) carbonate and dolomite is magnesium carbonate.

Properties of carbonates

Most metal carbonates thermally decompose when heated to form the metal oxide and carbon dioxide gas. For example:

copper(II) carbonate ⟶ copper(II) oxide + carbon dioxide

$$CuCO_3(s) \longrightarrow CuO(s) + CO_2(g)$$

Group 1 metal carbonates, except for lithium carbonate, do not dissociate on heating. It is found generally that the carbonates of the more reactive metals are much more difficult to dissociate than, for example, copper(II) carbonate.

Carbonates are generally insoluble in water except for those of sodium, potassium and ammonium.

Carbonates react with acids to form salts, carbon dioxide and water.

calcium carbonate + hydrochloric acid ⟶ calcium chloride + carbon dioxide + water

$$CaCO_3(s) + 2HCl(aq) \longrightarrow CaCl_2(aq) + CO_2(g) + H_2O(l)$$

This reaction is used in the laboratory preparation of carbon dioxide. It is also used as a test for a carbonate, because the reaction produces carbon dioxide which causes effervescence and if bubbled through limewater turns it chalky white.

Properties of hydrogencarbonates

Sodium hydrogencarbonate, unlike the carbonate, decomposes very readily on heating releasing carbon dioxide and water vapour.

sodium hydrogencarbonate ⟶ sodium carbonate + carbon dioxide + water

$$2NaHCO_3(aq) \longrightarrow Na_2CO_3(s) + CO_2(g) + H_2O(l)$$

Sodium hydrogencarbonate is often called bicarbonate of soda. It is used in baking powder, a raising agent in baking bread and cakes.

Quick Questions

1 Write word and chemical equations for the reaction of magnesium carbonate with dilute hydrochloric acid.
2 Describe the chemical test for carbon dioxide gas.
3 This question is about the limestone cycle.

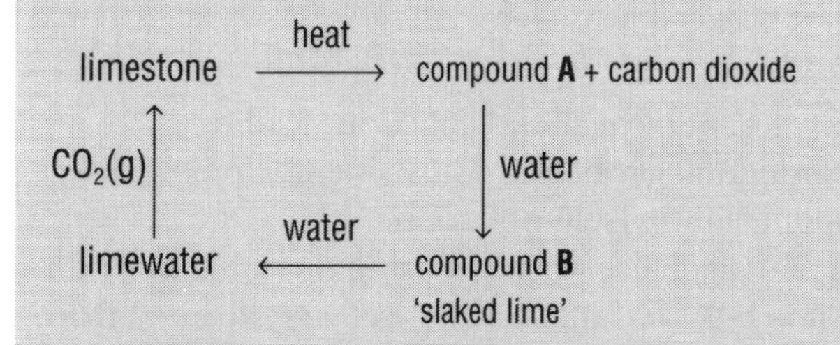

a) Name and give the formula of:
i) compound **A**; ii) compound **B**.
b) Write balanced chemical equations for the formation of both compounds **A** and **B**.
c) Name and give the symbol/formula for the ions present in limewater.
d) Describe with the aid of a balanced chemical equation what happens if carbon dioxide is bubbled through limewater.

3.5 Substances from oil

Fossil fuels

Coal, oil and natural gas are all examples of **fossil fuels**. The term fossil fuels is derived from the fact that they are formed from dead plants and animals which fossilised over 200 million years ago during the Carboniferous era.

Coal was produced by the action of pressure and heat on dead wood from ancient forests.

Thick forest grew in swamp land in many parts of the world. When dead trees fell into the swamps, they were buried by mud. The lack of oxygen prevented **aerobic** decay from taking place. Over millions of years, due to the Earth's movement as well as changes in climate, the land sank and the decaying wood became covered by even more layers of mud and sand.

Anaerobic decay occurred and as time passed, the gradually forming coal became more and more compressed as other material was laid down above it. Over millions of years as the layers of forming coal were pushed deeper and the pressure and temperature increased, the final conversion to coal took place.

Different types of coal were formed as a result of different pressures being applied during its formation. For example, anthracite is a much harder and higher carbon content coal, typical of coal produced at greater depths.

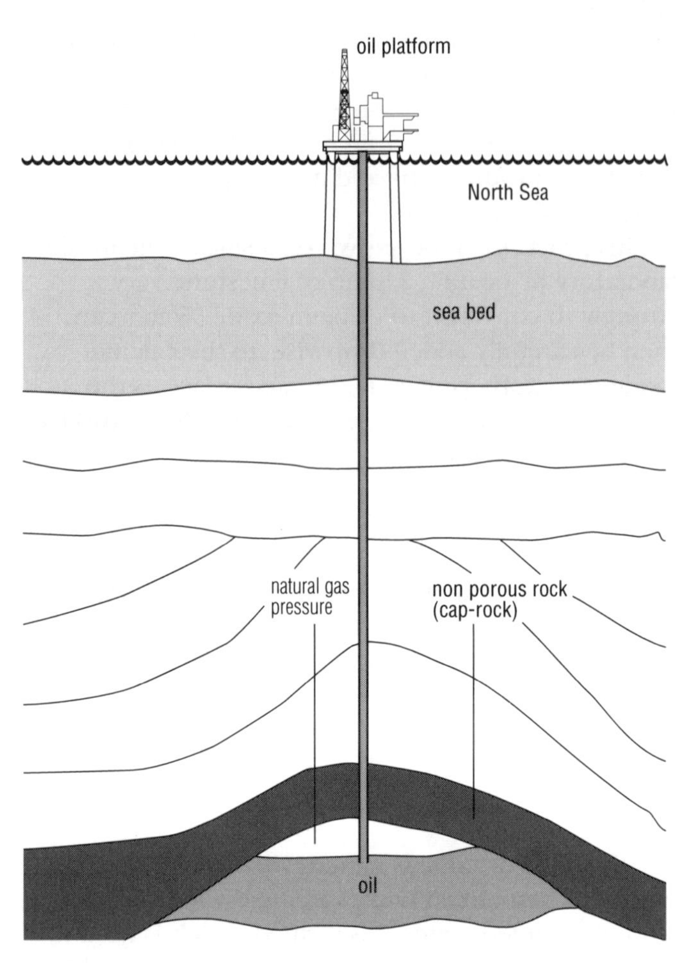

Figure 3.5.1 Natural gas and oil are trapped under non-porous rock

Table 3.5.1

Type of coal	Carbon content %
Anthracite	90
Bituminous coal	60
Lignite	40
Peat	20

The mining of peat and coal does lead to environmental problems. These include noise, dirt, subsidence and visual pollution.

Oil and gas were formed during the same period as coal. It is believed that oil and gas were formed from the remains of plants, animals and bacteria that once lived in seas and lakes. This material sank to the bottom of these seas and lakes and became covered in mud, sand and silt which thickened with time. Anaerobic decay took place and as the mud layers built up, high temperatures and pressures were created which converted the material slowly into oil and gas.

As rock formed, Earth movements caused it to buckle and split and the oil and gas was trapped in folds beneath layers of non-porous rock or cap rock.

Oil refining

Crude oil is a complex mixture of compounds known as **hydrocarbons**. Hydrocarbons are molecules which contain only the elements carbon and hydrogen, bonded together covalently. These carbon compounds form the basis of a group called **organic compounds**. Living things are all made from organic compounds. Crude oil is not only a major source of fuels but it is also a **raw material** of enormous importance. It supplies a large and diverse chemical industry which produces dozens of products from it.

Crude oil is not very useful to us until it has been processed. The process, known as refining, is carried out at an oil refinery.

Refining involves separating crude oil into various batches, or **fractions**, using a technique called fractional distillation. The different fractions separate because they have different boiling points. The crude oil is heated to about 400 °C to vaporise all the different parts in the mixture. The mixture of vapours passes into the fractionating column near the bottom. Each fraction is obtained by collecting hydrocarbon molecules which have a boiling point in a given range of temperatures.

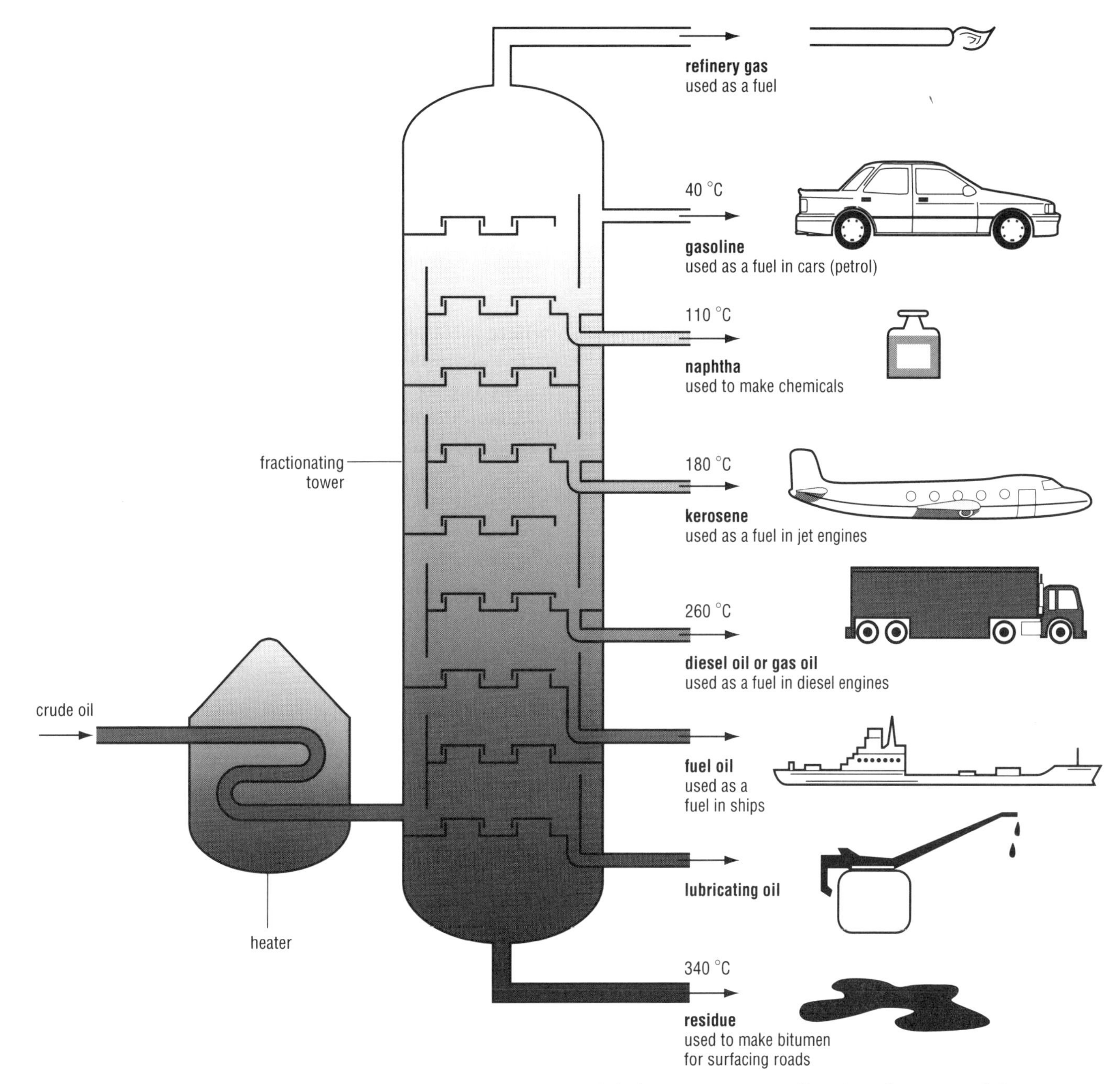

Figure 3.5.2 The fractional distillation of crude oil in a refinery

For example, the fraction we know as petrol contains molecules which have boiling points between 30 and 110 °C. The molecules in this fraction contain 5–10 carbon atoms. These smaller molecules with lower boiling points condense higher up the tower. The bigger hydrocarbon molecules, which have the higher boiling points, condense in the lower half of the tower.

The liquids which condense at different levels are collected on trays. These fractions usually contain a number of different hydrocarbons. The individual single hydrocarbon molecules can then be obtained by further refining the fraction by further distillation. We can duplicate this process, in the laboratory, by using the apparatus shown in Figure 1.5.6, Section 1.5.

It is important to realise that the uses of the **fractions** depends on their properties. For example, one of the higher fractions, which boils in the range 250–350 °C is quite thick and sticky and makes a good **lubricant**. However, the petrol fraction burns very easily and this therefore makes it a good fuel for use in petrol engines.

Quick Questions

1. What do you understand by the term hydrocarbon?
2. All organisms are composed of compounds which contain carbon. Why do you think carbon chemistry is often called organic chemistry?
3. List the main fractions obtained by separating the crude oil mixture and explain how they are obtained in a refinery.

3.6 Alkanes

Most of the hydrocarbons in crude oil belong to the family of compounds called **alkanes**. The molecules within the alkane family contain carbon atoms, covalently bonded to four other atoms by single bonds. These molecules possess only single covalent bonds. They are said to be **saturated** because no further atoms can be added.

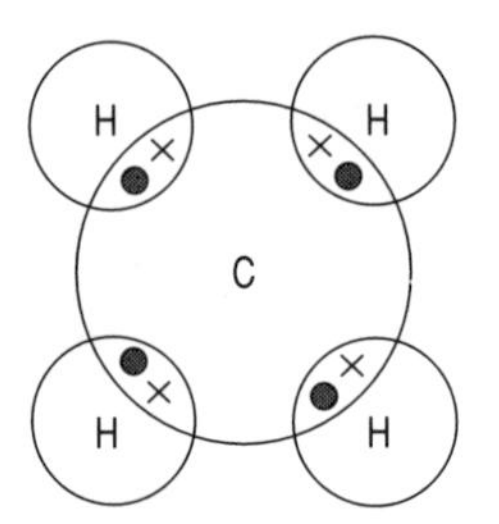

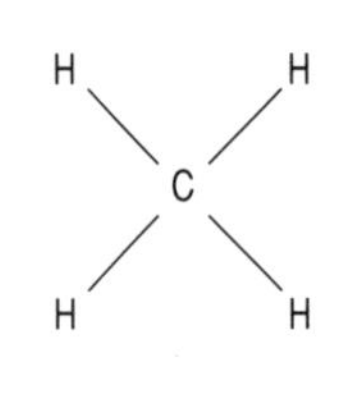

Figure 3.6.1

The first six members of the alkane family and their physical properties are shown in Table 3.6.1.

Table 3.6.1

Alkane	Formula	Melting point (°C)	Boiling point (°C)
Methane	CH_4	−182	−164
Ethane	C_2H_6	−183	−87
Propane	C_3H_8	−190	−42
Butane	C_4H_{10}	−138	0
Pentane	C_5H_{12}	−129	36
Hexane	C_6H_{14}	−95	69

The actual arrangement of the atoms in the first three members of this family is shown in Figure 3.6.2.

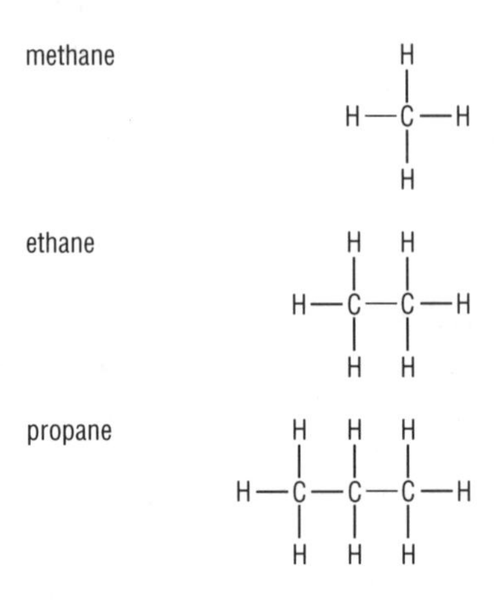

Figure 3.6.2

You will notice from the table above that the compounds have a similar structure and similar name endings (they all end in -ane).

They behave in a similar way chemically, and the family of compounds can be represented by a general formula. In the case of the alkanes the general formula is:

$$C_nH_{(2n+2)}$$

where n is the number of carbon atoms present.

For example, if $n = 2$ then the formula of the alkane is $C_2H_{(2\times2)+2} = C_2H_6$ (ethane).

A family with the above factors in common is called a **homologous series**.

As you go up a homologous series, in order of increasing number of carbon atoms, the physical properties of the compounds gradually change. For example, the melting and boiling points of the alkanes gradually increase. This is due to an increase in the intermolecular forces (van der Waals' forces) as the size of the molecule increases.

It is found that under normal conditions the molecules with up to four carbon atoms are gases, those molecules with between 5 and 16 carbon atoms are liquids, and those molecules with greater than 16 carbon atoms are solids.

Naming of the alkanes

All the alkanes have names ending -ane. The rest of the name tells you the number of carbon atoms present in the molecule. For example, the compound whose name begins with:

meth- has one carbon atom
eth- has two carbon atoms
prop- has three carbon atoms
but- has four carbon atoms
pent- has five carbon atoms
and so on.

Isomerism

Sometimes it is possible to write more than one **structural formula** to represent a **molecular formula**. The structural formula of a compound shows how the atoms are joined together by covalent bonds. For example, there are two different compounds with the molecular formula C_4H_{10}.

a butane

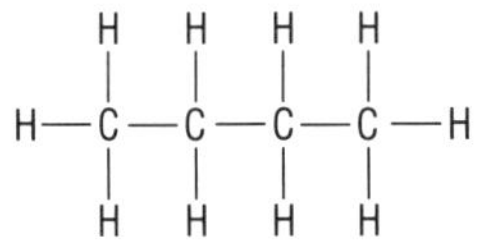

```
     H   H   H   H
     |   |   |   |
 H — C — C — C — C — H
     |   |   |   |
     H   H   H   H
```

melting point −138 °C
boiling point 0 °C

a 2–methylpropane

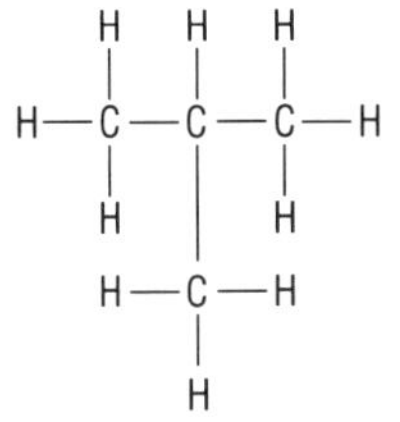

```
     H   H   H
     |   |   |
 H — C — C — C — H
     |   |   |
     H   |   H
     H — C — H
         |
         H
```

melting point −159 °C
boiling point −12 °C

Figure 3.6.3 The isomers of C_4H_{10}

Compounds that have the same molecular formula but different structural formulae are known as **isomers**. The different structures of the compounds shown have different melting and boiling points. 2-methylpropane contains a branched chain and has a lower melting point than butane, which has no branched chain.

Straight chained alkanes have more intermolecular forces between their molecules than branched chained molecules of the same molecular formula. This is because the branching makes the molecule more spherical, reducing the surface area over which the intermolecular forces can operate.

All the alkane molecules with four or more carbon atoms possess isomers.

The chemical behaviour of alkanes

The alkanes are rather unreactive compounds. For example, they are generally not affected by alkalis, acids or many other substances. Their most important property is that they burn easily.

The gaseous alkanes such as methane will burn in a good supply of air, forming carbon dioxide and water as well as plenty of heat energy.

$$\text{methane} + \text{oxygen} \longrightarrow \text{carbon dioxide} + \text{water} + \text{energy}$$

$$CH_4(g) + 2O_2(g) \longrightarrow CO_2(g) + 2H_2O(g)$$

The gaseous alkanes are some of the most useful fuels. Methane, better known as natural gas, is used for cooking as well as heating our offices, schools and homes.

Propane and butane burn with very hot flames and they are sold as the liquefied petroleum gases (LPG), Propagas and Calor gas respectively. In rural areas where there is no supply of natural gas, central heating systems can be run on propane gas. Butane, sometimes mixed with propane, is used in portable blow lamps and for gas lighters.

Quick Questions

1. Name the alkanes which have the following formulae:
   a) C_8H_{18};
   b) $C_{10}H_{22}$;
   c) C_6H_{14}.
2. Using the information given in Table 3.6.1, estimate the boiling points of:
   a) C_7H_{16};
   b) C_9H_{20}.
3. Draw structural formulae for:
   a) C_8H_{18};
   b) $C_{10}H_{22}$.
4. Draw the structural formulae of the isomers of:
   a) C_5H_{12};
   b) C_6H_{14}.
5. Write a word and balanced chemical equation to represent the combustion of:
   a) Ethane;
   b) Propane.
6. Draw dot and cross diagrams to show the covalent bonding present in:
   a) Ethane;
   b) Butane.

3.7 Alkenes

This is another homologous series of hydrocarbons of general formula:

$$C_nH_{2n}$$

where n is the number of carbon atoms.

The **alkenes** are more reactive than the alkanes because they each contain a **double covalent bond** between the carbon atoms.

```
H       H
 \     /
  C = C
 /     \
H       H
```

Figure 3.7.1 The covalent bonding in ethene, the simplest alkene

Molecules which possess a double covalent bond of this kind are said to be **unsaturated** because it is possible to break this double bond and add extra atoms to the molecule.

The names of all the alkenes end in -ene. The alkenes, especially ethene, are very important industrial chemicals. They are used extensively in the plastics industry and in the production of alcohols such as ethanol and propanol.

Table 3.7.1

Alkene	Formula	Melting point (°C)	Boiling point (°C)
Ethene	C_2H_4	−169	−104
Propene	C_3H_6	−185	−47
Butene	C_4H_8	−184	−6

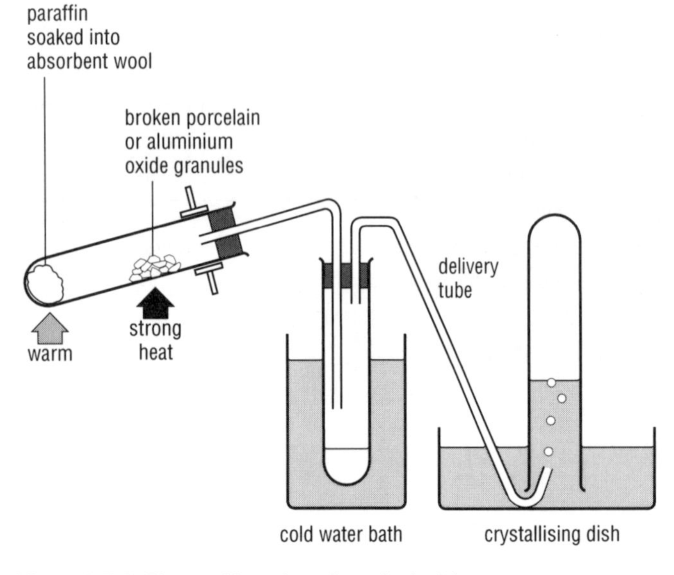

Figure 3.7.2 The cracking of an alkane in the laboratory

Where do we get alkenes from?

Very few alkenes are found in nature. Most are obtained by breaking up larger, less useful alkane molecules. This is usually done by a process called **catalytic cracking**. In this process the alkane molecules to be 'cracked' (split up) are passed over a mixture of aluminium and chromium oxides heated to about 500 °C.

dodecane	⟶	decane	+	ethene
$C_{12}H_{26}(g)$ (found in kerosene)	⟶	$C_{10}H_{22}(g)$ shorter alkane	+	$C_2H_4(g)$ alkene

In the laboratory, a catalyst of broken, unglazed pottery is used.

The chemical behaviour of alkenes

The double bond makes the alkenes more reactive than the alkanes, since it tends to break open during chemical reactions and join onto other atoms. These are known as **addition reactions**. For example, under suitable conditions hydrogen will 'add' across the double bond, forming ethane.

$$\text{ethene} + \text{hydrogen} \xrightarrow[\text{Ni or Pt catalyst}]{150\text{–}300\ ^\circ\text{C}} \text{ethane}$$

$$C_2H_4(g) + H_2(g) \longrightarrow C_2H_6(g)$$

```
H       H                   H   H
 \     /                    |   |
  C = C     + H—H  ⟶   H—C—C—H
 /     \                    |   |
H       H                   H   H
```

This reaction is called hydrogenation. Hydrogenation reactions like the one shown with ethene are used in the manufacture of margarines from vegetable oils. Vegetable oils contain fatty acids, such as linoleic acid ($C_{18}H_{32}O_2$).

These are unsaturated molecules, containing several double bonds. These double bonds make the molecule less flexible. By hydrogenation, it is possible to convert these molecules into more saturated ones. Now the molecules are less rigid and can flex and twist more easily, and hence pack more closely together. This in turn causes an increase in the intermolecular forces and so raises the melting point. The now solid margarines can be spread on bread more manageably than liquid oils.

Another important addition reaction is the one used in the manufacture of ethanol. Ethanol has important uses as a solvent and a fuel. It is formed when water (as steam) is added across the double bond in ethene. For this reaction to take place, the reactants have to be passed over a catalyst of phosphoric(V) acid (absorbed on silica pellets) at a temperature of 300 °C and pressure of 60 atmospheres (1 atmosphere = 1×10^5 Pascals).

$$\text{ethene} + \text{steam} \underset{\text{phosphoric(V) acid catalyst}}{\overset{300\ ^\circ\text{C, 60 Atm}}{\rightleftharpoons}} \text{ethanol}$$

$$C_2H_4(g) + H_2O(g) \rightleftharpoons C_2H_5OH(g)$$

$$H_2C{=}CH_2 + H{-}OH \rightleftharpoons H{-}CH_2{-}CH_2(OH){-}H$$

The $\rightleftharpoons$ sign shows this reaction is **reversible**. To ensure the highest possible yield of ethanol, the conditions have been chosen which favour the forward reaction.

A test for unsaturated compounds

The addition reaction between bromine dissolved in an organic solvent e.g. 1,1,1-trichloroethane and alkenes is used as a chemical test for the presence of a double bond between two carbon atoms. A few drops of this bromine solution are shaken with the hydrocarbons. If it is an alkene, such as ethene, a reaction takes place in which bromine joins to the alkene double bond. This results in the bromine solution losing its colour. If an alkane, such as hexane, is shaken with the bromine solution then no colour change takes place. This is because there are no double bonds present between the carbon atoms of alkanes.

$$\text{ethene} + \text{bromine} \longrightarrow \text{dibromoethane}$$

$$C_2H_4(g) + Br_2(aq) \longrightarrow C_2H_4Br_2(aq)$$

$$H_2C{=}CH_2 + Br{-}Br \longrightarrow H{-}CH(Br){-}CH(Br){-}H$$

Quick Questions

1 Make an estimate of the boiling point of hexene.
2 Write a balanced chemical equation to represent the process that takes place when decane is cracked.
3 Describe, with the aid of a diagram, what is meant by the term addition reaction.
4 Which of the following formulae represent alkanes and which represent alkenes?
C_6H_{12}; C_5H_{12}; C_9H_{20}; $C_{12}H_{24}$; $C_{20}H_{42}$; C_2H_4.
5 Draw a dot and cross diagram to show the covalent bonding in:
a) Ethene;
b) Propene.

3.8 Polymers

Polymers consist of very large molecules, each one made up of many thousands of smaller molecules called **monomers**. The word polymer is derived from the Greek 'poly' meaning many and the '-mer' from the word monomer, therefore polymer, many monomers.

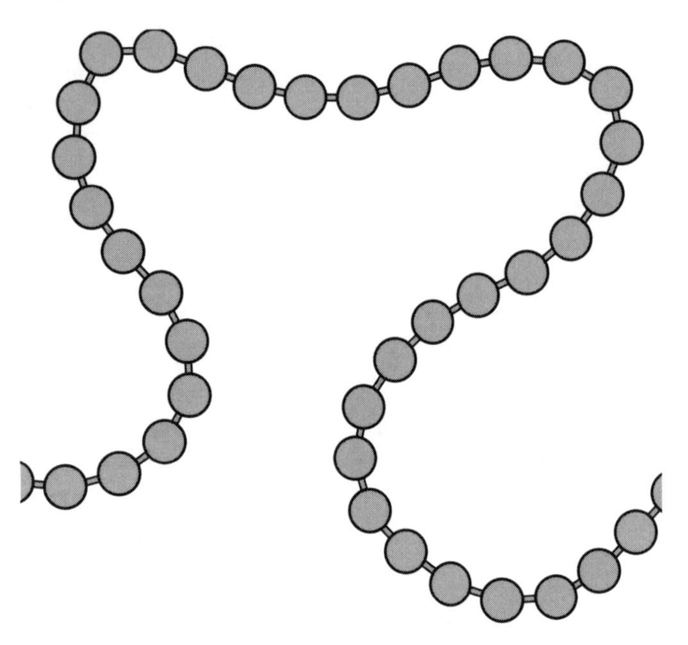

Figure 3.8.1 Molecules of ethene in this polythene chain are represented by poppet beads

A very common, and useful, polymer is poly(ethene) or polythene. Polythene has many useful properties:

- It is easily moulded.
- It is an excellent electrical insulator.
- It does not corrode.
- It is tough.
- It is not affected by the weather.
- It is durable.

Its properties allow it to be used as a substitute for natural materials in things such as bags, sandwich boxes, washing up bowls, wrapping film, milk bottle crates and squeezy bottles.

Polythene is manufactured by heating ethene to a relatively high temperature, under a high pressure in the presence of a catalyst.

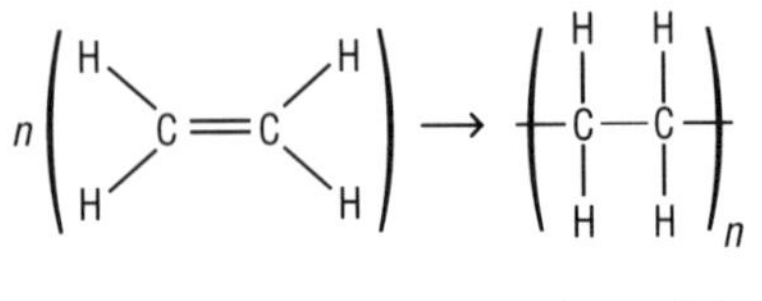

$$n\left(\begin{matrix} H & & H \\ & C{=}C & \\ H & & H \end{matrix}\right) \rightarrow \left(\begin{matrix} & H & H & \\ - & C & - C & - \\ & H & H & \end{matrix}\right)_n$$

monomer — polymer chain

where n is a very large number

Figure 3.8.2 In polythene, the ethene molecules have joined together to form a very long hydrocarbon chain

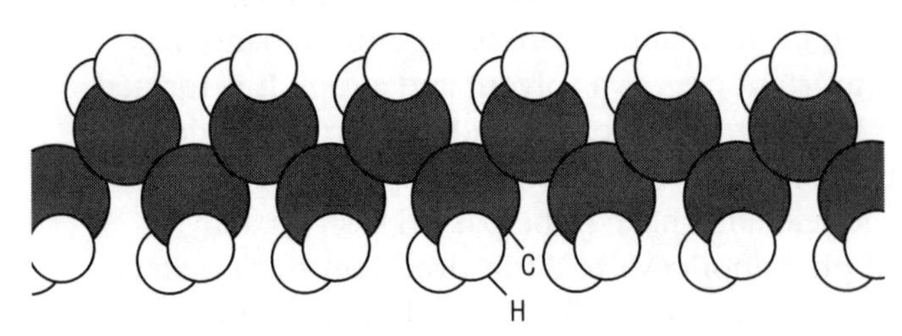

Figure 3.8.3 This shows part of the polythene polymer chain

The ethene molecules are able to form chains like this because they possess carbon–carbon double bonds. Other alkene molecules can also produce substances like polythene. For example, propene produces polypropene which is used to make ropes and packaging.

When small molecules like ethene join together to form long chains of atoms, the process is called **polymerisation**. A polymer chain often consists of many thousands of monomer units and in any piece of plastic there will be many millions of polymer chains. Since in this polymerisation process the monomer units add together to form the polymer, the process is called **addition polymerisation**.

Other addition polymers

Many other addition polymers have been produced, for example, PTFE (polytetrafluoroethene) and PVC (polyvinyl chloride or polychloroethene). Both of these plastics have monomer units similar to ethene.

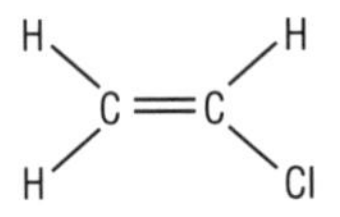

$$\begin{matrix} H & & H \\ & C{=}C & \\ H & & Cl \end{matrix} \qquad\qquad \begin{matrix} F & & F \\ & C{=}C & \\ F & & F \end{matrix}$$

PVC monomer (vinyl chloride or chloroethene) 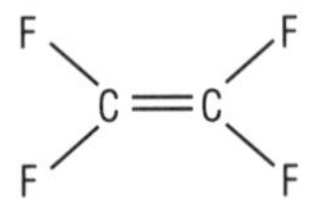 — PTFE monomer (tetrafluoroethene)

Often, the plastics are produced with particular properties in mind.

If we start from chloroethene, the polymer we make is slightly stronger and harder than polythene and is therefore particularly good for making pipes for plumbing.

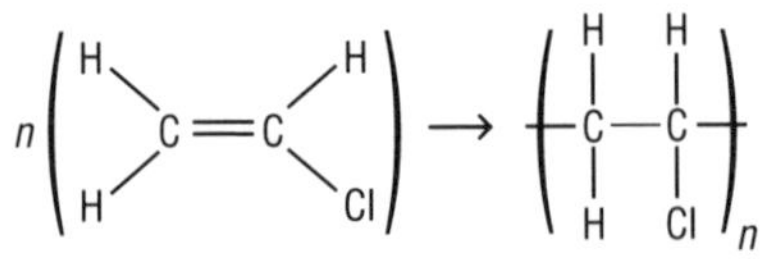

$$n\left(\begin{matrix} H & & H \\ & C{=}C & \\ H & & Cl \end{matrix}\right) \rightarrow \left(\begin{matrix} & H & H & \\ - & C & - C & - \\ & H & Cl & \end{matrix}\right)_n$$

monomer — polymer chain

If we start from tetrafluoroethene, the polymer we make has some slightly unusual properties:

- It will stand very high temperatures.
- It forms a very slippery surface.

These properties make PTFE an ideal 'non-stick' coating for frying pans.

$$n\,(F_2C{=}CF_2) \rightarrow (-CF_2-CF_2-)_n$$

monomer — polymer chain

Table 3.8.1

Plastic	Monomer	Properties	Uses
Polyethene	$H_2C{=}CH_2$	Tough, durable	Carrier bags, bowls, buckets, packaging
Polypropene	$CH_3(H)C{=}CH_2$	Tough, durable	Ropes, packaging
PVC	$H_2C{=}C(H)Cl$	Strong, hard (less flexible than polythene)	Pipes, electrical insulation, guttering
PTFE	$F_2C{=}CF_2$	Non-stick surface, withstands high temperaure	Non-stick frying pans, soles of irons
Polystyrene	$C_6H_5(H)C{=}CH_2$	Light, poor conductor of heat	Insulation, packaging, (especially as foam)
Perspex	$H_2C{=}C(CO_2CH_3)CH_3$	Transparent	Used as a glass substitute

Disposal of plastics

In the last thirty years plastics have taken over as replacement materials for metals, glass, paper and wood as well as for natural clothing fibres such as cotton and wool. This is not surprising since they are light, cheap, relatively unreactive, colourful and can be easily moulded. However this situation has created a waste problem of enormous proportions.

In the recent past, much of our plastic waste has been used to landfill disused quarries but these sites are getting harder to find and it is becoming more and more expensive.

The alternatives to dumping plastic waste certainly are more economical and more satisfactory. For example:

- Incineration – schemes have been developed to use the heat generated for heating purposes.
- Recycling – large quantities of black plastic bags and sheeting are produced for resale.
- Biodegradable plastics – Plastics have been developed which break down under natural conditions. These include plastics which are broken down by bacteria, plastics which dissolve in water and those which decompose in sunlight.

Quick Questions

1. Write the general equation to represent the formation of polystyrene from its monomer.
2. Suggest other uses for the polymers shown in Table 3.8.1.
3. Give two advantages and two disadvantages of plastic waste (rubbish).

3.9 Alcohols and carboxylic acids

Alcohols

The **alcohols** are another homologous series with general formula $C_nH_{2n+1}OH$.

For example, if $n = 2$ then $C_2H_{(2\times2)+1}OH = C_2H_5OH$ - ethanol. All the alcohols possess an $-OH$ as the **functional group**. The functional group is the group of atoms responsible for the characteristic reactions of an organic compound.

Table 3.9.1

Alcohol	Formula	Melting point (°C)	Boiling point (°C)
Methanol	CH_3OH	−94	64
Ethanol	CH_3CH_2OH	−117	78
Propanol	$CH_3CH_2CH_2OH$	−126	97

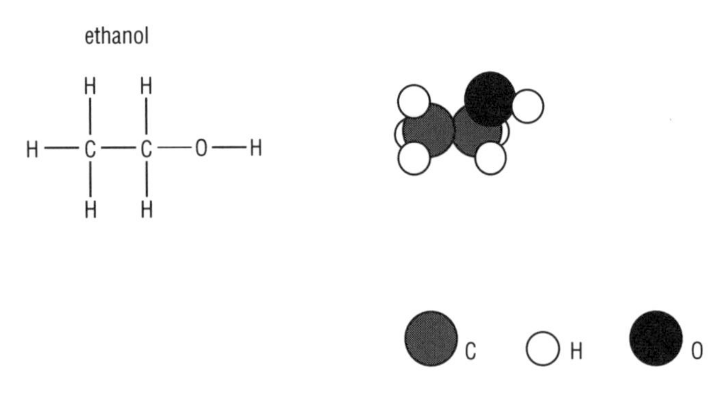

Figure 3.9.1 The bonding in ethanol

Ethanol is by far the most important of the alcohols and is usually just called alcohol. It is a neutral, colourless, volatile liquid which does not conduct electricity. Ethanol burns quite readily with a clean, hot flame.

ethanol + oxygen ⟶ carbon dioxide + water + energy

$$CH_3CH_2OH(l) + 3O_2(g) \longrightarrow 2CO_2(g) + 3H_2O(g)$$

As methylated spirit, or meths, it is used in spirit (camping) stoves. Methylated spirit is ethanol to which small amounts of poisonous substances have been added to stop you drinking it! Also, some countries including Brazil use ethanol as a fuel for cars. Up to 20% of ethanol can be added to petrol used in car engines without the need to make adjustments to the carburettor.

Many other materials such as food flavourings are made from ethanol. Ethanol is a very good solvent and evaporates easily. It is therefore used extensively as a solvent for paints, glues, after shave and many other everyday products.

Almost all the world's supply of ethanol is produced by the hydration of ethene. However, some is produced by **fermentation**. It involves a series of biochemical reactions brought about by micro-organisms or enzymes. Fermentation is the basic process behind the baking, wine and beer making industries.

Fermentation in the laboratory can be carried out using sugar solution. A micro-organism called **yeast** is added to the solution. The yeast uses the sugar for energy during anaerobic respiration (respiration without air). In the process of doing this, the sugar is broken down to give carbon dioxide and ethanol. The best temperature for this process to be carried out at is 37 °C.

glucose $\xrightarrow{\text{yeast}}$ ethanol + carbon dioxide

$$C_6H_{12}O_6(aq) \longrightarrow 2C_2H_5OH(l) + 2CO_2(g)$$

Alcoholic drinks like beer and wine are made on the large scale in vast quantities. Beer is made from barley to which hops have been added to produce the distinctive flavour, whilst wine is made by fermenting grape juice which contains glucose. The micro-organisms in yeast will carry out the fermentation quite successfully in both cases.

Beer normally contains only about 4% by volume of ethanol, whilst wine contains about 11% by volume of ethanol. The yeast is killed off if there is much more ethanol than this. So it is not possible to make stronger drinks by fermentation. Some of the stronger drinks in Table 3.9.2, like sherry, are made up to strength by adding pure ethanol.

Table 3.9.2

Drink	% ethanol by volume
Whisky, brandy	40
Sherry	20
Martini	14
Wine	12
Beer	4

Carboxylic acids

Ethanol can be oxidised to ethanoic acid (an organic acid also called acetic acid) by powerful oxidising agents such as warm acidified potassium dichromate(VI). During the reaction the orange colour of potassium dichromate(VI) changes to a dark green as the ethanol is oxidised to ethanoic acid.

$$\text{ethanol} + \underset{\text{from potassium dichromate(VI)}}{[\text{oxygen}]} \xrightarrow{\text{heat}} \text{ethanoic acid} + \text{water}$$

$$CH_3CH_2OH(l) + 2[O] \longrightarrow CH_3COOH(aq) + H_2O(l)$$

A similar oxidation process takes place, but more slowly, if wine or beer is left open. It will eventually turn to 'vinegar' (the sharp taste of ethanoic acid) due to bacterial oxidation of the ethanol in the alcoholic drink.

carboxylic acid group

Figure 3.9.2 The bonding in ethanoic acid

Ethanoic acid is a member of another homologous series of compounds called the **carboxylic acids**.

They have the general formula $C_nH_{2n+1}COOH$ and possess —COOH as their functional group.

Ethanoic acid behaves as a typical dilute acid and will react with bases, alkalis, metals and carbonates to produce salts called ethanoates.

Ethanoic acid will react with ethanol, in the presence of a few drops of concentrated sulphuric acid, to produce ethyl ethanoate - an **ester**.

$$\text{ethanoic acid} + \text{ethanol} \underset{}{\overset{\text{concentrated sulphuric acid}}{\rightleftharpoons}} \text{ethyl ethanoate} + \text{water}$$

$$CH_3COOH(l) + CH_3CH_2OH(l) \rightleftharpoons CH_3COOCH_2CH_3(aq) + H_2O(l)$$

Figure 3.9.3 Ethyl ethanoate

This reaction is called an esterification. Members of the ester family have strong and pleasant smells. Many of them occur naturally and are responsible for the flavours in fruits and the smells of flowers. They are used, therefore, in some food flavourings and in perfumes.

Fats and oils are naturally occurring esters which are used as energy storage compounds by plants and animals.

Quick Questions

1 Write the structural formula for propanol.

2 Write a word and balanced chemical equation for:
   a) The combustion of propanol;
   b) The oxidation of propanol.

3 As you will know, ethanol (alcohol) is a drug and is addictive. Write a paragraph to explain some of the problems associated with alcohol.

4 Write:
   a) The structural formula for propanoic acid;
   b) A word and balanced chemical equation for the esterification of propanoic acid with ethanol.

5 The diagram shows the arrangement of the outer electrons only in a molecule of ethanoic acid.

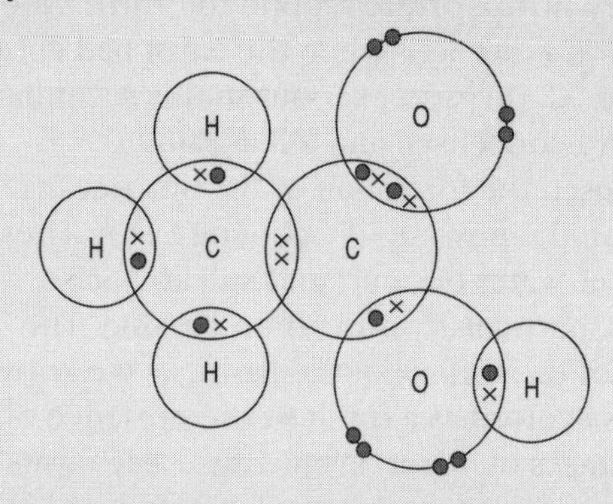

   a) Name the different elements found in this compound.
   b) What is the total number of atoms present in this molecule?
   c) Between which two atoms is there a double covalent bond?
   d) How many single covalent bonds does each carbon atom have?
   e) Write a paragraph explaining the sorts of properties you would expect this sort of substance to have.

6 Explain why when wine is left open it eventually turns to vinegar.

7 **a)** Copy the following table and complete it by writing the structural formulae for methanol and methanoic acid.

Methane	Methanol	Methanoic acid
H H—C—H H		

   b) Describe a simple chemical test that could be used to distinguish methanol from methanoic acid.
   c) i) Name the class of compound produced when methanol reacts with methanoic acid.
   ii) Name the type of reaction taking place.
   iii) Write a word and balanced chemical equation for this reaction.
   iv) Give two uses related to the class of compound formed in this reaction.

3.10 The developing atmosphere

About 4500 million years ago the **planets** in the solar system were formed.

Each planet had a thick layer of gases, mainly hydrogen and helium, surrounding its core, known as the **primary atmosphere**. Over a period of time, intense solar activity caused these lighter gases to be removed from the planets nearest to the Sun. During this time, the Earth was cooling to become a molten mass upon which a thin crust formed.

Volcanic activity through the crust pushed out huge quantities of gases such as ammonia, nitrogen, methane, carbon monoxide, carbon dioxide and a small amount of sulphur dioxide which formed a **secondary atmosphere** around the Earth. About 3800 million years ago, when the Earth had cooled below 100 °C, the water vapour in this secondary atmosphere condensed and fell as rain.

This caused the formation of the first oceans, lakes and seas on the now rapidly cooling Earth. These expanses of water became, and still are, 'gas reservoirs' particularly for carbon dioxide. The structure of the surface of the Earth, as we know it today, has evolved as a result of the presence of these large expanses of water. Eventually, early forms of life developed in these oceans, lakes and seas at depths which prevented potentially harmful ultraviolet light from the Sun affecting them. About 3000 million years ago primitive 'algae' like plants appeared along with the first forms of bacteria. These algae used the light from the Sun, via **photosynthesis**, to produce their own food and oxygen was released into the atmosphere as a waste product. The process of photosynthesis can be described by the following equation:

$$\text{carbon dioxide} + \text{water} \xrightarrow[\text{chlorophyll}]{\text{sunlight}} \text{glucose} + \text{oxygen}$$

$$6CO_2(g) + 6H_2O(l) \longrightarrow C_6H_{12}O_6(aq) + 6O_2(g)$$

The **ultraviolet radiation** broke down some of the oxygen molecules, which had been released, into oxygen atoms.

$$\text{oxygen molecules} \xrightarrow{\text{uv light}} \text{oxygen atoms}$$

$$O_2(g) \longrightarrow 2O(g)$$

Some of the highly reactive oxygen atoms reacted with molecules of oxygen to form ozone molecules, $O_3(g)$.

$$\text{oxygen atom} + \text{oxygen molecule} \longrightarrow \text{ozone}$$

$$O(g) + O_2(g) \longrightarrow O_3(g)$$

Ozone is an unstable molecule which readily decomposes under the action of ultraviolet radiation to form single oxygen atoms and oxygen molecules.

$$\text{ozone} \xrightarrow{\text{uv light}} \text{oxygen atom} + \text{oxygen molecule}$$

$$O_3(g) \longrightarrow O(g) + O_2(g)$$

However, the single oxygen atoms react quickly – reforming ozone molecules.

Ozone is an important gas in the atmosphere (stratosphere, see Figure 3.10.1). It prevents harmful ultraviolet radiation from reaching the Earth. Over many millions of years, the amount of ozone increased so the amount of ultra violet radiation was reduced significantly. About 400 million years ago the first land plants appeared on the Earth and so the amount of oxygen and hence ozone increased further.

Oxygen is a reactive gas and over millions of years organisms adapted to make use of it. The oxygen from the atmosphere was used, along with the carbon they obtained from their food, to produce energy in a process known as **respiration**.

The process of respiration can be represented as:

$$\text{glucose} + \text{oxygen} \longrightarrow \text{carbon dioxide} + \text{water} + \text{energy}$$

$$C_6H_{12}O_6(aq) + 6O_2(g) \longrightarrow 6CO_2(g) + 6H_2O(l) + \text{energy}$$

About 350 million years ago, simple animals began to develop which did not rely on sunlight for their energy.

Structure of the atmosphere

The gases in the atmosphere are held in an envelope around the Earth by its gravity. The atmosphere is 80 km thick and it is divided into five layers.

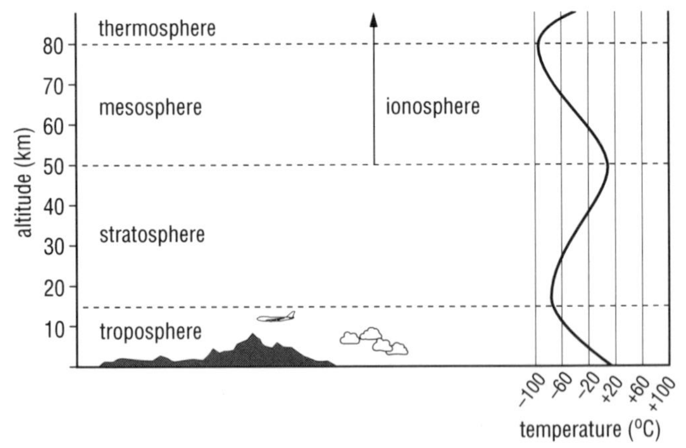

Figure 3.10.1 The Earth's atmosphere

About 75% of the mass of the atmosphere is found in the layer nearest the Earth, which is called the troposphere. Beyond the troposphere the atmosphere reaches into space but becomes extremely thin beyond the mesosphere.

The composition of the atmosphere

If a sample of dry, unpolluted air was taken from anywhere in the troposphere and analysed, the composition by volume of the sample would be similar to that shown in the table below.

Table 3.10.1

Component	%
Nitrogen	78.08
Oxygen	20.95
Argon	0.93
Carbon dioxide	0.03
Neon	0.002
Helium	0.0005
Krypton	0.0001
Xenon	0.00001 plus minute amounts of other gases

Measuring the percentage of oxygen in the air

When 100 cm^3 of air is passed backwards and forwards over the heated copper turnings it is found that the amount of gas decreases. This is because the reactive part of the air, the oxygen gas, is reacting with the copper to form black copper(II) oxide.

copper + oxygen ⟶ copper(II) oxide

$$2Cu(s) + O_2(g) \longrightarrow 2CuO(s)$$

In such an experiment, the volume of gas in the syringe decreases from 100 cm^3 to about 79 cm^3, showing that the air contained 21 cm^3 of oxygen gas.

The percentage of oxygen gas in the air is:

$$\frac{21}{100} \times 100 = 21\%$$

Quick Questions

1. Recently, 'holes' have been observed in the ozone layer. Why is this a potential problem?
2. The apparatus shown in Figure 3.10.2, was used to estimate the proportion of oxygen in the atmosphere.
   A volume of dry air (90 cm^3) was passed backwards and forwards over heated copper until no further change in volume took place. The apparatus was then allowed to cool down to room temperature and the final volume was taken. The results are shown below.
   Volume of gas before passing over heated copper = 90.0 cm^3.
   Volume of gas after passing over heated copper = 70.7 cm^3.
   a) Why was the apparatus allowed to cool back to room temperature before the final volume reading was taken?
   b) Using the information given, calculate the volume reduction and hence the percentage reduction in the volume.
   c) Explain briefly why there was a change in volume.
   d) Give the name of the main residual gas at the end of the experiment.

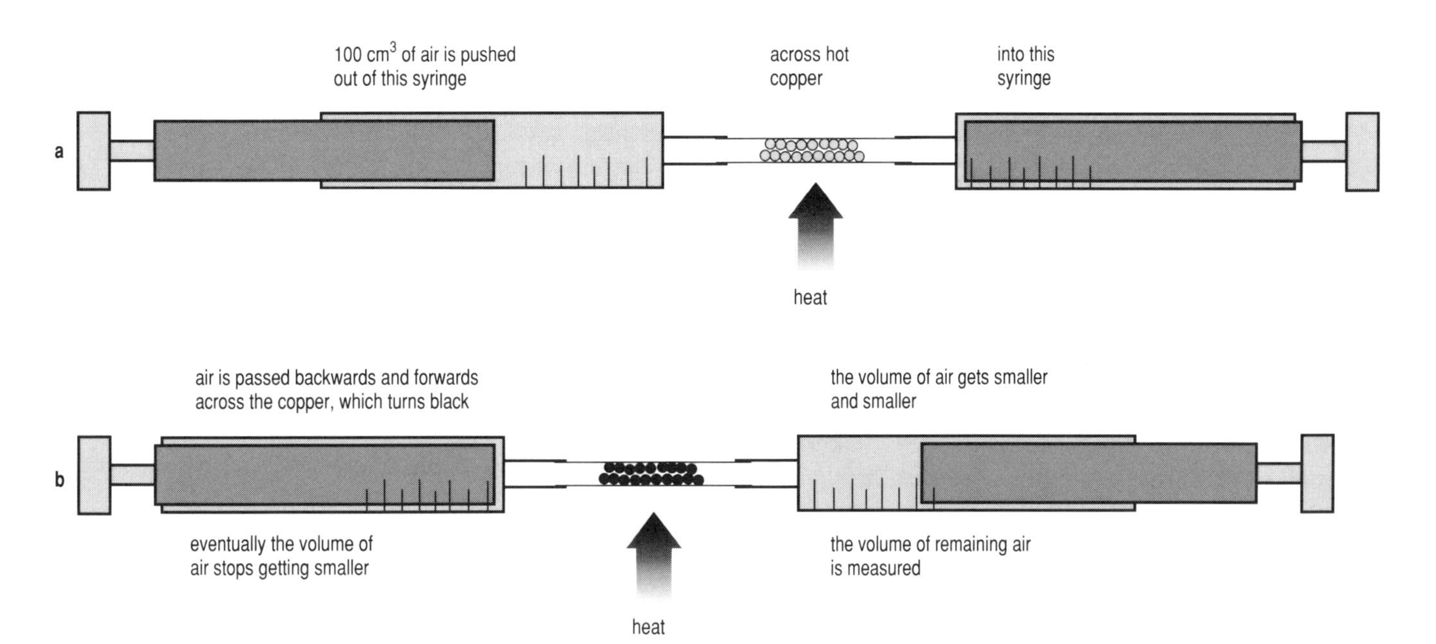

Figure 3.10.2 This apparatus can be used to find out the volume of oxygen gas in the air

3.11 The changing atmosphere

The composition of the atmosphere is affected by respiration, photosynthesis, volcanic activity as well as the products of radioactive decay. It is also affected by human activity, involving burning of fossil fuels.

Carbon dioxide

Carbon forms two oxides, carbon monoxide (CO) and carbon dioxide (CO_2). Carbon dioxide is the more important of the two and in industry large amounts are obtained from the liquefaction of air. Air contains approximately 0.03% by volume of carbon dioxide. This value has remained almost constant for a long period of time and is maintained via the carbon cycle, see Figure 3.11.1. However, recently scientists have detected a slight increase in the amount of carbon dioxide in the atmosphere.

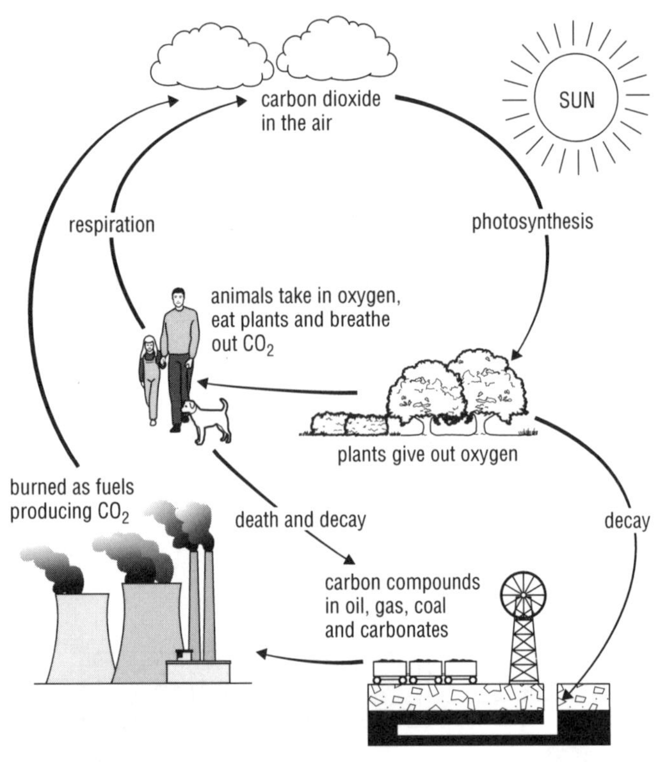

Figure 3.11.1 The carbon cycle

Carbon dioxide is produced by burning fossil fuels. It is also produced by all living organisms through aerobic respiration.

This carbon dioxide is taken in by plants through their leaves and used together with water, taken in through their roots, to synthesise sugars.

This cycle has continued in this manner for millions of years. The effect which results is known as the **greenhouse effect**. The greenhouse effect is needed for life on Earth. It is the increase in the greenhouse effect (global warming) that causes problems.

Carbon dioxide in the Earth's atmosphere allows the Sun's rays through onto the surface of the Earth, but stops heat from escaping. This effect is similar to that which is observed in a greenhouse where sunlight (visible and ultraviolet radiation) enters through the glass panes, but heat (infrared radiation) has difficulty in escaping through the glass. More carbon dioxide in the atmosphere means more heat is trapped in, causing temperatures to rise.

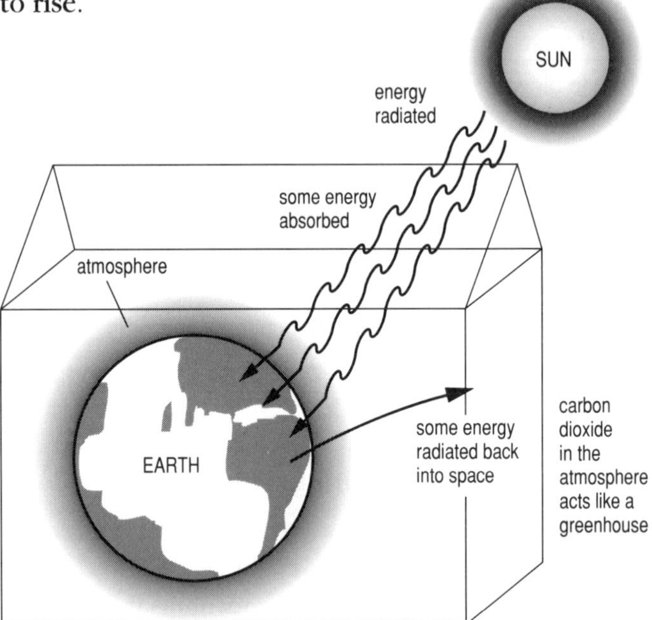

Figure 3.11.2 The greenhouse effect

The Earth's climate is affected by the levels of carbon dioxide (and water vapour) in the atmosphere. If the amount of, in particular, carbon dioxide builds up in the air it is thought that the average temperature of the Earth will rise. Due to the increase in the amount of carbon dioxide produced through burning fossil fuels and the deforestation of large areas of tropical rain forest (so less CO_2 is absorbed in photosynthesis), scientists have detected an imbalance in the carbon cycle.

The long term effect of the higher temperatures will be the gradual melting of ice caps and consequent flooding in low lying areas of the Earth. There will also be changes in the weather patterns of the Earth which would affect agriculture worldwide.

Uses of carbon dioxide

- Fizzy drinks are made by dissolving carbon dioxide gas in water under pressure.
- It is used in fire extinguishers for use on electrical fires. Carbon dioxide is denser than air and forms a layer around the burning material. It covers the fire and starves it of oxygen.

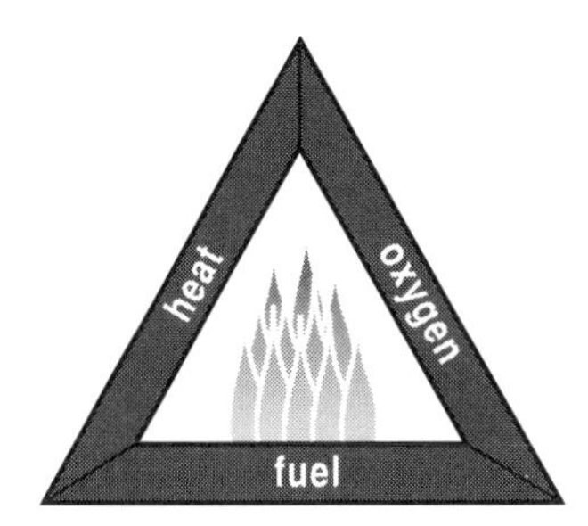

Figure 3.11.3 The fire triangle

- Solid carbon dioxide (dry ice) is used for refrigerating ice cream, meat and soft fruits.
- Solid carbon dioxide is used to create the smoke effect you see at pop concerts and on television. Dry ice is placed in boiling water and it forms thick clouds of white 'smoke'. It stays close to the floor due to the fact that carbon dioxide is denser than air.
- Carbon dioxide gas is also used for transferring heat in some nuclear power stations.

Laboratory preparation of carbon dioxide gas

In the laboratory the gas is made by pouring dilute hydrochloric acid onto marble chips ($CaCO_3$).

calcium carbonate + hydrochloric acid ⟶ calcium chloride + water + carbon dioxide

$$CaCO_3(s) + 2HCl(aq) \longrightarrow CaCl_2(aq) + H_2O(l) + CO_2(g)$$

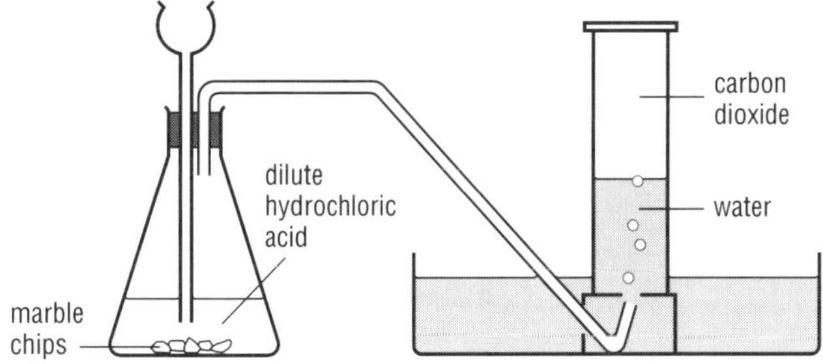

Figure 3.11.4 Apparatus for preparing CO_2

Properties of carbon dioxide gas

Carbon dioxide has the following physical properties. It is a colourless gas, sparingly soluble in water as well as being denser than air.

Chemical properties

When bubbled into water it dissolves slightly and some of the carbon dioxide reacts, forming a solution of the weak acid carbonic acid. This has a pH of 4 or 5.

water + carbon dioxide ⇌ carbonic acid

$$H_2O(l) + CO_2(g) \rightleftharpoons H_2CO_3(aq)$$

It will only support the combustion of strongly burning substances such as magnesium. This burning reactive metal decomposes the carbon dioxide to provide oxygen for its continued burning in the gas.

magnesium + carbon dioxide ⟶ magnesium oxide + carbon

$$2Mg(s) + CO_2(g) \longrightarrow 2MgO(s) + C(s)$$

When carbon dioxide is bubbled through limewater (calcium hydroxide solution) a white precipitate is formed. The white solid is calcium carbonate ($CaCO_3$). This is used as a test to show whether a gas is carbon dioxide.

carbon dioxide + calcium hydroxide ⟶ calcium carbonate + water

$$CO_2(g) + Ca(OH)_2(aq) \longrightarrow CaCO_3(s) + H_2O(l)$$

If carbon dioxide is bubbled through this solution continuously, it will eventually become clear. This is because of the formation of calcium hydrogencarbonate which is soluble in water.

calcium carbonate + water + carbon dioxide ⟶ calcium hydrogencarbonate

$$CaCO_3(s) + H_2O(l) + CO_2(g) \longrightarrow Ca(HCO_3)_2(aq)$$

Atmospheric pollution

Air pollution is all around us. Gases such as carbon monoxide, sulphur dioxide and nitrogen oxides are increasing in concentration in the atmosphere as the population increases, with the consequent rise in the need for energy, industries and motor vehicles. These gases are produced primarily from the combustion of the fossil fuels coal, oil and gas but they are also produced by the smoking of cigarettes.

Power stations are a major source of sulphur dioxide, a pollutant which is formed by the combustion of coal, oil and gas which contain small amounts of sulphur.

sulphur + oxygen ⟶ sulphur dioxide

$$S(s) + O_2(g) \longrightarrow SO_2(g)$$

Sulphur dioxide gas dissolves in rain water forming the weak acid, sulphurous acid (H_2SO_3).

sulphur dioxide + water ⇌ sulphurous acid

$$SO_2(g) + H_2O(l) \rightleftharpoons H_2SO_3(aq)$$

A further reaction occurs in which the sulphurous acid is oxidised to sulphuric acid. Solutions of these acids are the principle contributors to **acid rain**.

Quick Questions

1. When carbon dioxide is 'poured' from a gas jar onto a burning candle, the candle goes out. What properties of carbon dioxide does this experiment demonstrate?
2. In the test for carbon dioxide, a precipitate of calcium carbonate is formed when the gas is bubbled through limewater. Write an ionic equation to show the formation of this precipitate.

3.12 Products from the air

Air is the major source of oxygen, nitrogen and the inert gases. The gases are obtained by fractional distillation of liquid air but it is a difficult process, involving several different steps.

- The air is passed through fine filters to remove any dust.
- The air is cooled to about −80 °C to remove water vapour and carbon dioxide as solids. If these are not removed then serious blockages of pipes can result.
- Next the cold air is compressed to about 100 atmospheres of pressure. This warms up the air, so it is passed into a heat exchanger to cool it down again.
- The cold, compressed air is allowed to expand rapidly, cooling it still further.
- The process of compression followed by expansion is repeated until the air reaches a temperature below −200 °C. At this temperature the majority of the air is liquefied see Table 3.12.1.

Table 3.12.1

Gas	Boiling point (°C)
Helium	−269
Neon	−246
Nitrogen	−196
Argon	−186
Oxygen	−183
Krypton	−157
Xenon	−108

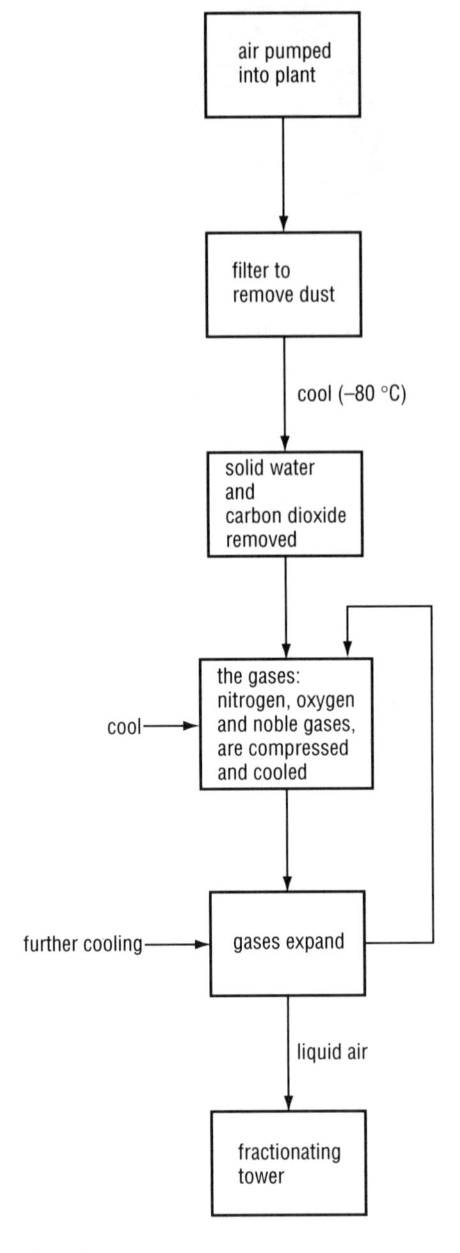

Figure 3.12.1

- The liquid air is passed into a fractionating column and it is fractionally distilled.
- The gases are then stored separately in large tanks and cylinders.

It should be noted that the inert gases neon, argon, krypton and xenon are obtained by this method, but helium is more profitably obtained from natural gas.

Compare the components of our atmosphere with those of the other planets in the solar system - see Table 3.12.2.

Table 3.12.2

Planet	Atmosphere
Mercury	No atmosphere – gases burned off by the heat of the Sun
Venus	Carbon dioxide and sulphur dioxide
Mars	Mainly carbon dioxide
Jupiter	Ammonia, helium, hydrogen, methane
Saturn	Ammonia, helium, hydrogen, methane
Uranus	Ammonia, helium, hydrogen, methane
Neptune	Helium, hydrogen, methane
Pluto	No atmosphere – it is frozen

Uses of the gases

Oxygen

- Large quantities are used in industry to convert pig iron into steel and for producing very hot flames for welding, by mixing with gases such as ethyne (acetylene).
- It is used in hospitals to help with breathing difficulties.
- Sportsmen such as mountaineers and divers use oxygen.
- It is carried in space rockets so that the hydrogen and kerosene fuels can burn.
- The space shuttle carried oxygen gas to use in fuel cells, for converting chemical energy into electrical energy.
- Astronauts must carry their own supply of oxygen, as do firemen.
- It is used to restore life to polluted lakes and rivers and in the treatment of sewage.

Nitrogen

- Nitrogen is used in very large quantities in the production of ammonia gas which is used to produce nitric acid. Nitric acid is used in the manufacture of dyes, explosives and fertilisers.
- Liquid nitrogen is used as a refrigerant. Its low temperature (−196 °C) makes it useful for the fast-freezing of food.
- Because of its unreactive nature, nitrogen is used as an inert atmosphere for some processes and chemical reactions. For example, empty oil tankers are filled with nitrogen to prevent fires.
- It is used in food packaging to keep the food fresh. Used in crisp packets, it also prevents the crisps being crushed.

The inert gases

Argon is used:

- To fill ordinary light bulbs to prevent the tungsten filament from reacting with oxygen in the air and forming the oxide.
- To provide an inert atmosphere in arc welding and in the production of titanium metal.

Neon is used:

- In advertising signs because it glows red when electricity is passed through it.
- In the helium-neon gas laser.
- In Geiger-Muller tubes which are used for the detection of radioactivity.

Helium is used:

- To provide an inert atmosphere for welding.
- As a coolant in nuclear reactors.
- With 20% oxygen, as a breathing gas used by deep sea divers.
- To inflate the tyres of large aircraft.
- To fill airships and weather balloons.
- In the helium-neon laser.
- In low temperature research because of its low boiling point.

Krypton and xenon are used:

- In lamps used in photographic flash units, stroboscopic lamps and those used in lighthouses.

Quick Questions

1 Explain the following:
   a) Air is a mixture of elements and compounds.
   b) The percentage of carbon dioxide in the atmosphere does not significantly vary from 0.03%.
   c) When liquid air has its temperature slowly raised from −270 °C, helium is the first gas to boil.

2 How does oxygen help to restore life to polluted lakes?

3 Why is it important to have nitrogen in fertilisers?

4 Why is helium needed to produce an inert atmosphere for welding?

5 Air is a raw material from which several useful substances can be separated. They are separated in the following process.
Dry and 'carbon dioxide free' air is cooled under pressure. Most of the gases liquefy as the temperature falls below −200 °C. The liquid mixture is separated by fractional distillation. The boiling points of the gases left in the air after removal of water vapour and carbon dioxide are given in Table 3.12.1.
   a) Why is the air dried and carbon dioxide removed before it is liquefied?
   b) Which of the gases will not become liquid at −200 °C?
   c) Which of the substances in the liquid mixture will be the first to change from liquid to gas as the temperature is slowly increased?
   d) Give a use for each of the gases shown in Table 3.12.1.
   e) Use the data given in Table 3.10.1 (page 93) to calculate the volume of each of the gases found in 1 dm^3 of air.

3.13 Rusting

After a period of time, objects made of iron or steel will become coated with rust. The **rusting** of iron wastes enormous amounts of money in the UK each year.

Rust is an orange-red powder consisting mainly of hydrated iron(III) oxide ($Fe_2O_3.xH_2O$). Both water and oxygen are essential for iron to rust. If one of these two substances is not present, then rusting will not take place. The rusting of iron is encouraged by salt.

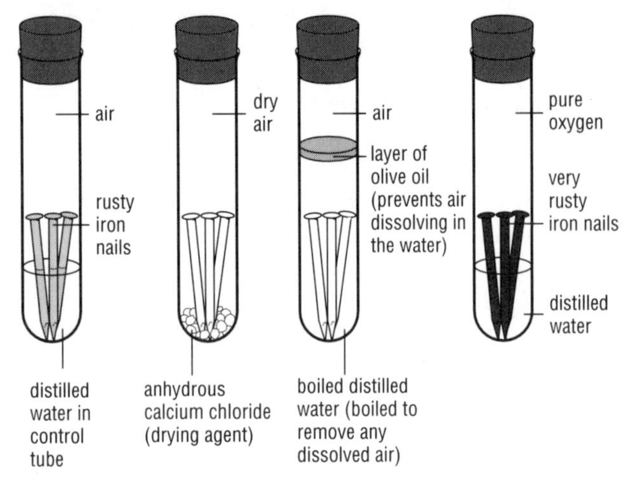

Figure 3.13.1 Rusting experiment with nails

Rust prevention

To prevent iron rusting, it is necessary to stop oxygen (from the air) and water coming into contact with it. There are several methods that can be used.

Painting

Ships, lorries, cars, bridges and many other iron and steel structures are painted to prevent rusting. If the paint, however, is scratched, the iron beneath it will start to rust and corrosion can then spread under the paintwork which is still sound. This is why it is essential that the paint is kept in good condition and checked regularly.

Oiling/greasing

The iron or steel in the moving parts of machinery are prevented from coming into contact with air or moisture by coating with oil. This is the most common method of protecting moving parts of machinery, but the protective film must be renewed.

Coating with plastic

The exteriors of refrigerators, freezers and many other items are coated with plastic, such as PVC, to prevent the steel structure rusting.

Plating

'Tin cans' for food are made from steel coated with tin. The tin is deposited onto the steel by electrolysis. Bicycle handle-bars are often plated with chromium to prevent rusting, giving a decorative finish.

Galvanising

Some steel girders, used in the construction of bridges and buildings, are galvanised. This involves dipping the girder into molten zinc. The thin layer of the more reactive zinc metal coating the iron slowly corrodes and loses electrons to the iron, so protecting it. This process continues even when much of the layer of zinc has been scratched away and so the iron continues to be protected.

Sacrificial protection

Bars of zinc are attached to the hulls of ships. Zinc is above iron in the reactivity series, it reacts in preference to it, and so the zinc is corroded and the iron or steel is protected. As long as some of the zinc bars remain in contact with the hull, the ship will be protected from rusting. When the zinc runs out, it must be renewed. Gas and water pipes made of iron and steel are connected by a wire to blocks of magnesium to obtain the same result. In both cases, as the more reactive metal corrodes, it loses electrons to the iron and so protects it (see Figure 3.13.2).

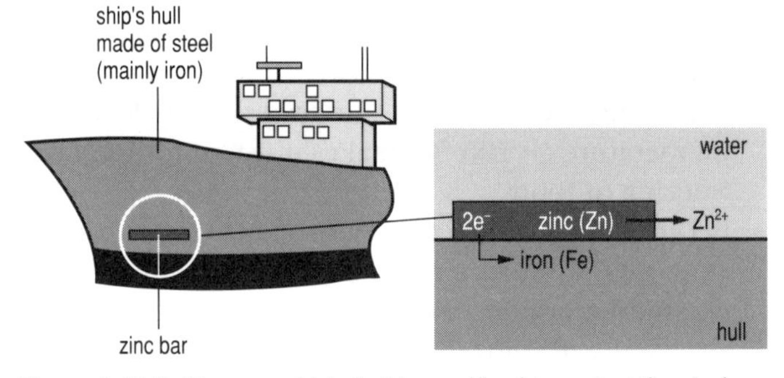

Figure 3.13.2 Zinc on a ship's hull is sacrificed to protect the steel

Corrosion

Rusting is the most common form of corrosion. Corrosion is the name given to the process which takes place when metals and alloys are chemically attacked by oxygen, water, or any other substances found in their immediate environment.

Generally, the higher the metal is in the reactivity series, the more rapidly it will corrode. If sodium and potassium were not stored under oil, they would corrode very rapidly indeed.

Magnesium, calcium and aluminium are usually covered by a thin coating of oxide after an initial reaction with oxygen in the air, which protects the metal from further oxidation. Freshly produced copper is pink but soon changes to brown due to the formation of copper(II) oxide on the surface of the metal.

In more exposed environments, copper roofs and pipes quickly become covered in verdigris. Verdigris is green and consists of copper salts formed on the surface of the copper. Gold and platinum are unreactive and do not corrode, even after hundreds of years.

Alloys

The majority of the metallic substances which are used today are **alloys**. Alloys are mixtures of two or more metals, and are formed by mixing molten metals thoroughly.

It is generally found that alloying produces a metallic substance with more useful properties than the original pure metals it was made from. Steel, which is a mixture of the metal, iron, and the non-metal, carbon, is also considered to be an alloy.

Of all the alloys we use, steel is perhaps the most important. Nickel and chromium are added to produce the alloy stainless steel. The chromium prevents the steel from rusting, whilst the nickel increases the hardness.

Production of steel

The 'pig iron' obtained from the blast furnace process contains about 5% of carbon and other impurities, including sulphur, silicon and phosphorus. These impurities make the iron hard and brittle. Most of them must be removed to improve the quality of the metal. In doing this, steel is produced. The impurities are removed in the **basic oxygen process**. In this process, molten iron from the blast furnace is poured into the basic oxygen furnace (see Figure 3.13.3).

A water-cooled lance is introduced into the furnace and oxygen at 5–15 atmospheres pressure is blown onto the surface of the molten metal. Carbon is oxidised to carbon monoxide and carbon dioxide, whilst sulphur is oxidised to sulphur dioxide. Silicon and phosphorus are oxidised to sulphur(IV) oxide and phosphorus pentoxide, which are solid oxides. Some calcium oxide (lime) is added to remove these solid oxides as slag. The slag may be skimmed or poured off the surface.

Samples are continuously taken and checked for carbon content. When the required amount of carbon has been reached, the blast of oxygen is turned off. The basic oxygen furnace can convert up to 300 tonnes of iron to steel every hour.

There are various types of steel which differ only in their carbon content.

If other types of steel are required, then up to 30% scrap steel, along with other metals (such as tungsten) are added and the carbon is burned off.

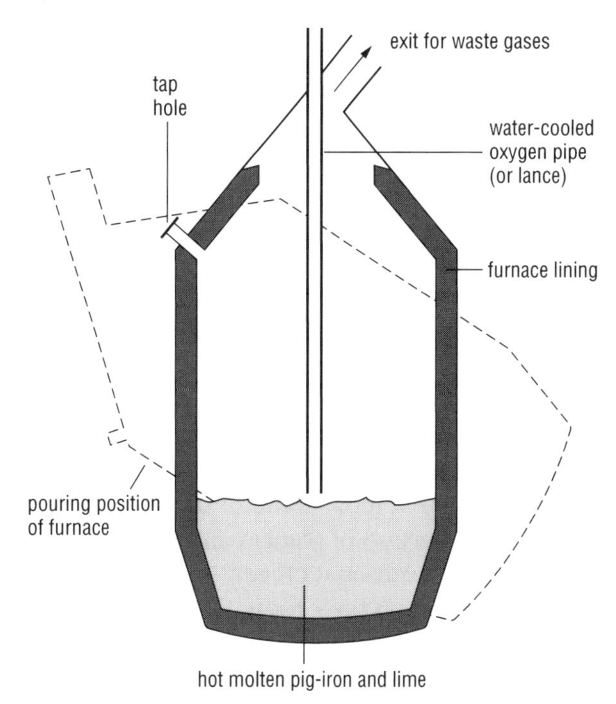

Figure 3.13.3 Basic oxygen furnace

Quick Questions

1. If the experiment shown in Figure 3.13.1 was left to run for two weeks, what changes would you expect to see on each nail after this time had elapsed?
2. How is the corrosion of aluminium useful to us?
3. Calcium oxide is a base. It combines with solid, acidic oxides in the basic oxygen furnace. Write a chemical equation for the reaction of one of these oxides with the added lime.
4. 'Many metals are more useful to us when mixed with some other elements.' Discuss this statement with respect to stainless steel.

Table 3.13.2 Metallurgists design alloys to do particular jobs

Alloy	Composition	Use
Brass	65% copper, 35% zinc	Jewellery, machine bearings, electrical connections, door furniture
Bronze	90% copper, 10% tin	Castings, machine parts
Cupro-nickel	30% copper, 70% nickel	Turbine blades
	75% copper, 25% nickel	Coinage metal
Duralumin	95% aluminium, 4% copper, 1% magnesium, manganese and iron	Aircraft construction, bicycle parts
Solder	70% lead, 30% tin	Connecting electrical wiring

3.14 Nitrogen

Nitrogen gas can be obtained industrially by the fractional distillation of liquid air. It is produced on a very large scale since the gas has many very important industrial uses.

Nitrogen is an essential element, necessary for the well-being of animals and plants. It is present in proteins which are found in all living things. Proteins are essential for healthy growth. Animals obtain the nitrogen they need for protein production by feeding on plants and other animals.

Most plants obtain the nitrogen they require from the soil. The nitrogen is found in compounds called nitrates which are produced as a result of the effects of lightning, and the decay of dead plants and animals. Nitrates are soluble compounds that can be absorbed by plants through their roots.

Nitrogen in the air is also converted into nitrates by some forms of bacteria. Some of these bacteria live in the soil. A group of plants called leguminous plants, including beans and clover, have these bacteria in nodules on their roots. These bacteria are able to take nitrogen from the atmosphere and convert it into a form in which it can be used to make proteins. This process is called **nitrogen fixation**.

The essential nature of nitrogen to both plants and animals can be summarised by the **nitrogen cycle**.

If farm crops are harvested from the land rather than left to decay, which would return the nitrogen to the soil, the soil becomes deficient in this important element. Nitrates can also be washed from the soil by the action of rain (leaching). For the soil to remain fertile for growing the next crop the nitrates need to be replaced.

Farmers often need to add substances which contain these nitrates. Such substances include:

- Farmyard manure, which is very rich in nitrogen-containing compounds that can be converted to nitrates by bacteria in the soil.
- **Artificial fertilisers.** These are manufactured compounds of nitrogen which are used on an extremely large scale to enable farmers to produce ever bigger harvests for the increasing population of the world. One of the most commonly used artificial fertilisers is ammonium nitrate, made from ammonia gas and nitric acid – both nitrogen containing compounds.

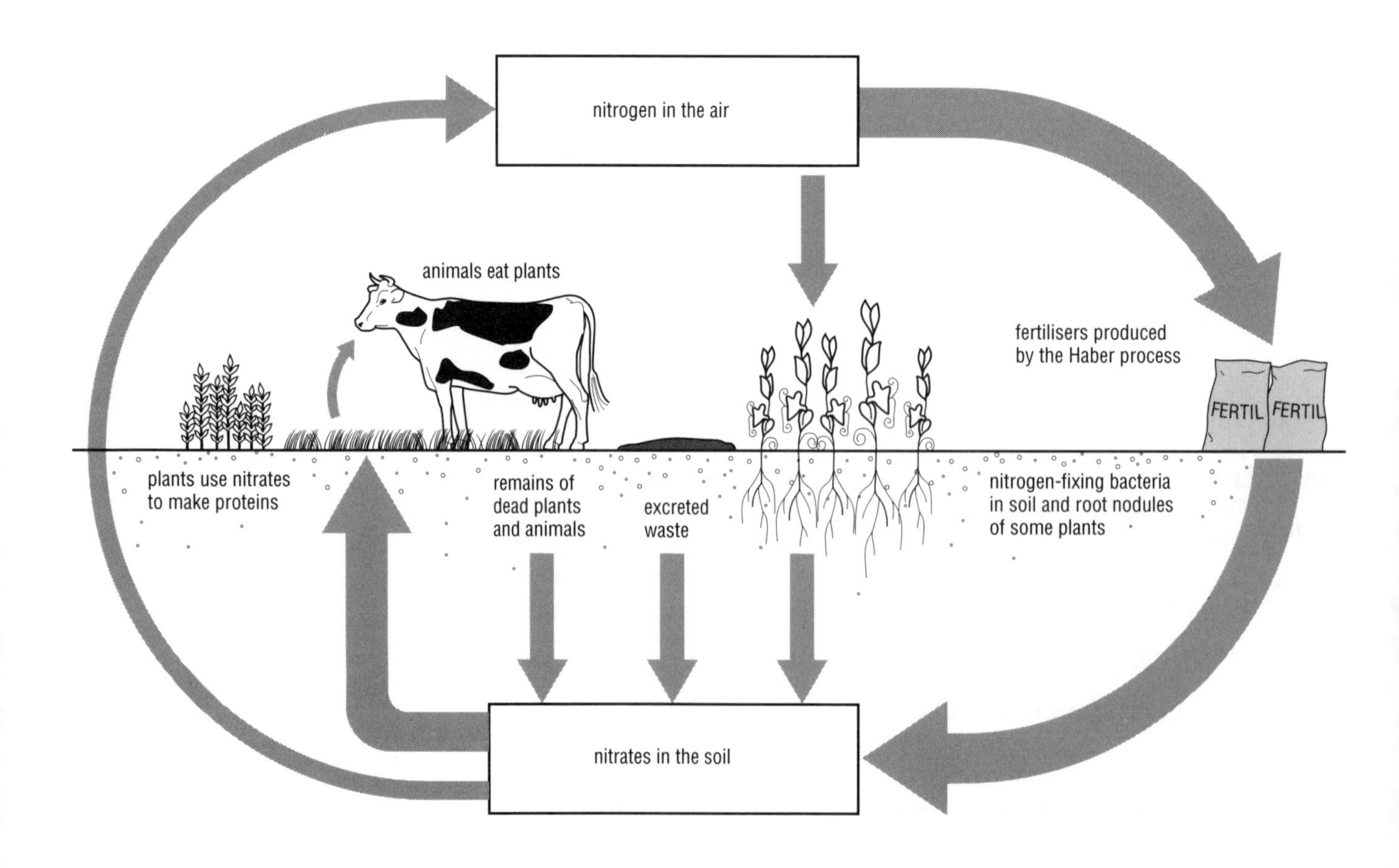

Figure 3.14.1 The nitrogen cycle

The Haber process

The **Haber process** is the basis of the artificial fertiliser industry as it involves the reaction of nitrogen and hydrogen to produce ammonia gas.

The nitrogen for the process is obtained from the atmosphere, and the hydrogen from the reaction between methane (natural gas) and steam.

methane + steam ⇌ hydrogen + carbon monoxide

$CH_4(g) + H_2O(l) \rightleftharpoons 3H_2(g) + CO(g)$

This process is known as steam reforming. This reaction is reversible, so the process is carried out at a temperature of 750 °C, a pressure of 30 atmospheres and using a nickel catalyst to ensure the reaction moves to the right (the forward reaction). These conditions are required to enable the maximum amount of hydrogen to be produced at an economic cost.

The carbon monoxide produced is then allowed to oxidise some of the unreacted steam to produce more hydrogen gas.

carbon monoxide + steam ⇌ hydrogen + carbon dioxide

$CO(g) + H_2O(l) \rightleftharpoons H_2(g) + CO_2(g)$

Nitrogen and hydrogen in the correct proportions (1 : 3) are then pressurised to approximately 200 atmospheres and passed over a catalyst of freshly produced, finely divided iron at a temperature of between 350 °C and 500 °C.

nitrogen + hydrogen ⇌ ammonia

$N_2(g) + 3H_2(g) \rightleftharpoons 2NH_3(g) \quad \Delta H = -92$ kJ/mol

Ammonia is produced in an exothermic reaction. The addition of hydrogen to a substance is an example of **reduction**. So in the above reaction the nitrogen is being reduced. The reverse of this process is **oxidation**.

Under these conditions, the gas mixture leaving the reaction vessel contains about 15% ammonia which is removed by cooling and condensing it as a liquid. The unreacted nitrogen and hydrogen are recirculated into the reaction vessel to react together once more to produce further quantities of ammonia.

The 15% production of ammonia does not seem much. The reason for this low percentage is the reversible nature of the reaction. Once the ammonia is formed, it decomposes to produce nitrogen and hydrogen. There comes a point when the rate at which the nitrogen and hydrogen react to produce ammonia is equal to the rate at which the ammonia decomposes. This situation is called a **chemical equilibrium**. Because the processes do not stop, the equilibrium is a **dynamic equilibrium**. The conditions used are chosen to ensure that the ammonia is made economically.

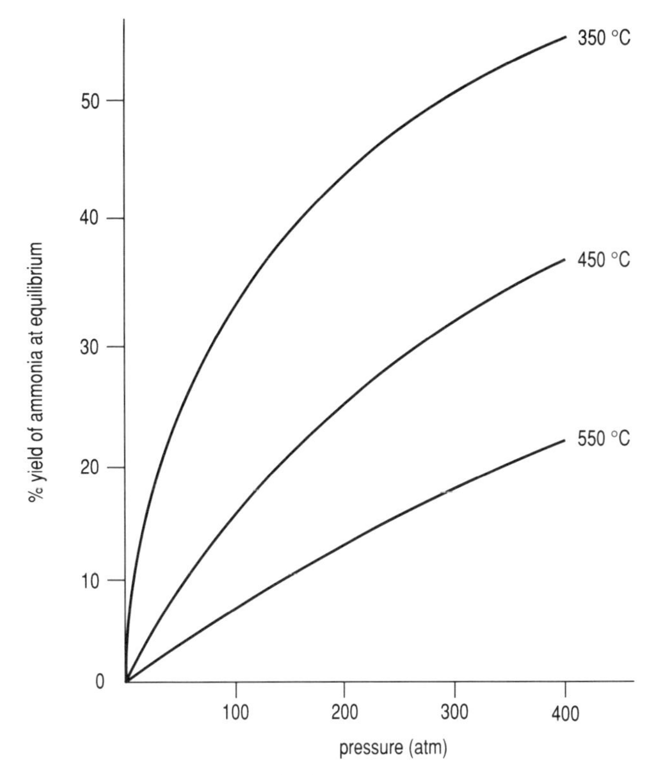

Figure 3.14.2 Yields from the Haber process

You will notice that the higher the pressure and lower the temperature used, the more ammonia is produced.

Relationships such as this were initially observed by Henri Le Chatelier, a French scientist, in 1888. He noticed that if the pressure was increased in reactions involving gases, the reaction producing the fewest molecules of gas was favoured. If you look at the reaction for the Haber process you will see that in going from left to right, the number of molecules of gas goes from four to two. This is why the Haber process is carried out at high pressures.

He also noticed that exothermic reactions produced more products if the temperature was low. Indeed, if the Haber process is carried out at room temperature, you do get a higher percentage of ammonia. However, in practice the rate of the reaction is lowered too much and the ammonia is not produced quickly enough for the process to be economic. So an **optimum temperature** is used, which produces enough ammonia at an acceptable rate. It should be noted, however, that the increased pressure used is very expensive in capital terms and so alternative, less expensive routes involving biotechnology are currently being sought.

Ammonia gas

Making ammonia in a laboratory

Small quantities of ammonia gas can be produced by heating any ammonium salt, such as ammonium chloride, with an alkali, such as sodium hydroxide.

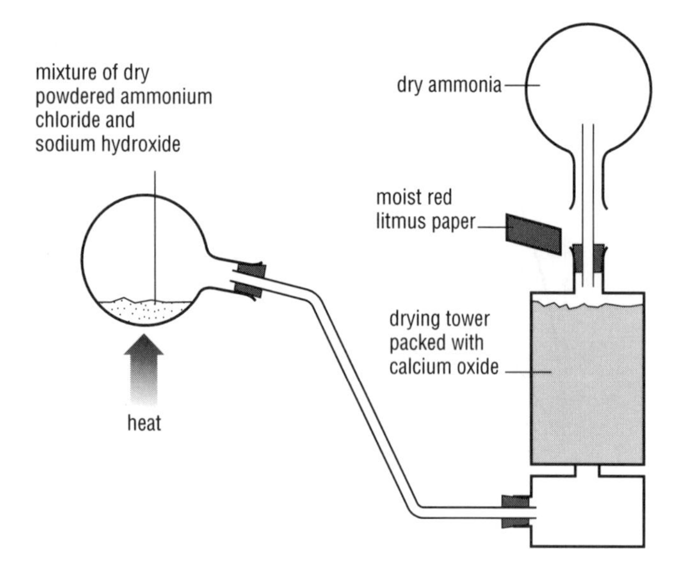

Figure 3.14.3

sodium hydroxide + ammonium chloride ⟶ sodium chloride + water + ammonia

$$NaOH(s) + NH_4Cl(s) \longrightarrow NaCl(s) + H_2O(g) + NH_3(g)$$

Water vapour is removed from the ammonia gas by passing the gas formed through a drying tower containing calcium oxide. This reaction forms the basis of a chemical test to show that a compound contains the ammonium ion (NH_4^+). If any compound containing the ammonium ion is heated with sodium hydroxide, ammonia gas will be given off. This will turn damp red litmus paper blue.

Physical properties of ammonia

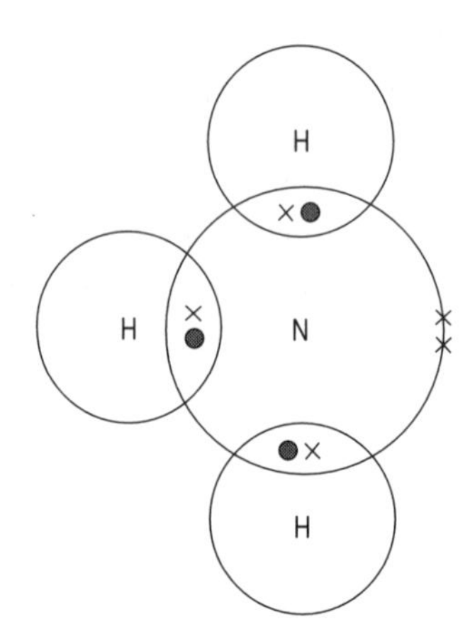

Figure 3.14.4 The ammonia molecule

Ammonia:

- Is a colourless gas.
- Is less dense than air.
- Has a sharp or pungent smell.
- Is very soluble in water, with about 1200 cm^3 of ammonia dissolving in each 1 cm^3 of water.

Chemical properties of ammonia

The reason ammonia is so soluble in water is that some of it reacts with the water. The high solubility can be shown by the fountain flask experiment.

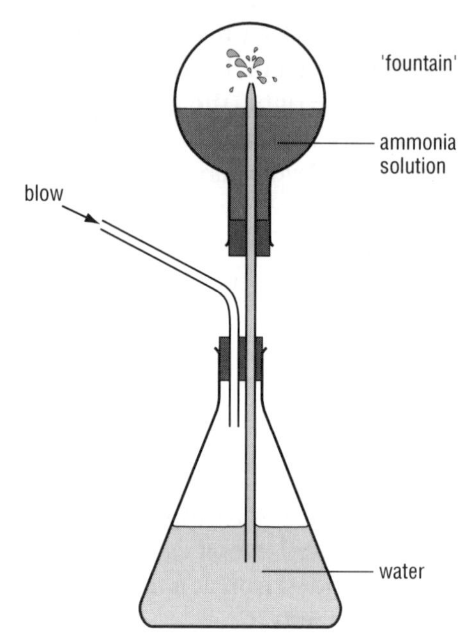

Figure 3.14.5 The fountain flask experiment

As the first drop of water reaches the top of the tube, all the ammonia gas in the flask dissolves – creating a much reduced pressure. Water then rushes up the tube to fill the space previously occupied by the now dissolved gas. This creates the fountain. If the water initially contained some universal indicator you would also see a colour change from green to blue when it comes into contact with the dissolved ammonia. This shows that ammonia solution is a weak alkali, although dry ammonia gas is not.

The reason for this is that a little of the ammonia gas has reacted with the water to produce ammonium ions and hydroxide ions. The hydroxide ions produced make a solution of ammonia alkaline.

ammonia + water ⇌ ammonium ions + hydroxide ions

$$NH_3(g) + H_2O(l) \rightleftharpoons NH_4^+(aq) + OH^-(aq)$$

The solution is only weakly alkaline because of the reversible nature of this reaction which results in a relatively low concentration of hydroxide ions. Ammonia gas dissolved in water is usually known as aqueous ammonia.

Ammonia solution will also form hydroxide precipitates with solutions containing certain metal ions. For example, it will give an orange/brown precipitate when added to a solution containing iron(III) ions, and a green precipitate when added to a solution of iron(II) ions. When added to a solution of copper ions, it initially forms a blue precipitate of copper(II) hydroxide which dissolves in excess ammonia to produce a deep blue solution.

Manufacture of nitric acid

One of the largest uses of ammonia is to produce nitric acid. The process, known as the Ostwald process occurs in three stages:

1 A mixture of air and ammonia is heated to about 230 °C and is passed through a metal gauze made of platinum (90%) and rhodium (10%). The reaction produces a lot of heat energy. This energy is used to keep the reaction vessel temperature at around 800 °C. The reaction produces nitrogen monoxide (NO) and water.

$$\text{ammonia} + \text{oxygen} \longrightarrow \text{nitrogen monoxide} + \text{water}$$

$$4NH_3(g) + 5O_2(g) \longrightarrow 4NO(g) + 6H_2O(g) \quad \Delta H = -904 \text{ kJ/mol}$$

2 The colourless nitrogen monoxide gas produced from the first stage is then reacted with oxygen from the air to form brown nitrogen dioxide gas (NO_2).

$$\text{nitrogen monoxide} + \text{oxygen} \longrightarrow \text{nitrogen dioxide}$$

$$2NO(g) + O_2(g) \longrightarrow 2NO_2(g) \quad \Delta H = -114 \text{ kJ/mol}$$

3 The nitrogen dioxide is then dissolved in water to produce nitric acid.

$$\text{nitrogen dioxide} + \text{water} \longrightarrow \text{nitric acid} + \text{nitrogen monoxide}$$

$$3NO_2(g) + H_2O(l) \longrightarrow 2HNO_3(aq) + NO(g)$$

Not all the nitrogen dioxide dissolves to form nitric acid. This is instead expelled into the atmosphere through a tall chimney. This can cause pollution because it is an acidic gas.

The nitric acid produced is used in the manufacture of:

- Artificial fertilisers such as ammonium nitrate.
- Explosives such as 2,4,6-trinitrotoluene (T.N.T.).
- Dyes.
- Man-made fibres such as nylon.
- Treatment of metals.

Artificial fertilisers

The two processes so far explained, the production of ammonia and nitric acid, are extremely important in the production of many artificial fertilisers.

These fertilisers are added to soil to replace nitrogen, phosphorus and potassium and other nutrients such as calcium, magnesium, sodium, sulphur, copper and iron. Examples of **nitrogenous fertilisers** (those which contain nitrogen) are shown in Table 3.14.1.

Table 3.14.1

Fertiliser	Formula
Ammonium nitrate	NH_4NO_3
Ammonium phosphate	$(NH_4)_3PO_4$
Ammonium sulphate	$(NH_4)_2SO_4$
Urea	$CO(NH_2)_2$

Artificial fertilisers can also enable land which was once unable to support crop growth to become fertile.

The fertilisers which add the three main nutrients (N, P and K) are called NPK fertilisers. They contain ammonium nitrate (NH_4NO_3), ammonium phosphate ($(NH_4)_3PO_4$) and potassium chloride (KCl) in varying proportions.

Manufacture of ammonium nitrate

Ammonium nitrate (Nitram©) is probably the most widely used nitrogenous fertiliser. It is manufactured by reacting ammonia gas and nitric acid.

$$\text{ammonia} + \text{nitric acid} \longrightarrow \text{ammonium nitrate}$$

$$NH_3(g) + HNO_3(aq) \longrightarrow NH_4NO_3(aq)$$

Problems with fertilisers

Problems can arise from the incorrect use of artificial fertilisers. If too much fertiliser is applied, rain washes it off the land and into rivers and streams. This process of leaching encourages the growth of algae and marine plants. As the algae die and decay, oxygen is removed from the water leaving insufficient for fish and other organisms.

Quick Questions

1 Why can most plants not use nitrogen gas directly from the atmosphere?
2 Nitrogen is essential in both plants and animals to form which type of molecule? Explain why these molecules are essential to life.
3 Use the information given to produce a flow diagram of the Haber process. Indicate the flow of the gases, and write equations to show what happens at each stage.
4 What problems do the builders of a chemical plant to produce ammonia have to consider, when they start to build such a plant?

3.15 Sulphur

Sulphur is a non-metallic element which plays a very important role in the chemical industry. It is a yellow solid found in large quantities in various forms throughout the world.

It is found in metal ores such as copper pyrites ($CuFeS_2$) and zinc blende (ZnS), as well as in volcanic regions of the world. Natural gas and oil contain sulphur and its compounds. The majority of this sulphur is removed as it would cause environmental problems.

Sulphur obtained from these sources is known as recovered sulphur and it is an important source of the element. It is also found as elemental sulphur in sulphur beds in Poland, Russia and the U.S.A. These sulphur beds are typically 200 m below the ground. Sulphur from these beds is extracted using the Frasch process.

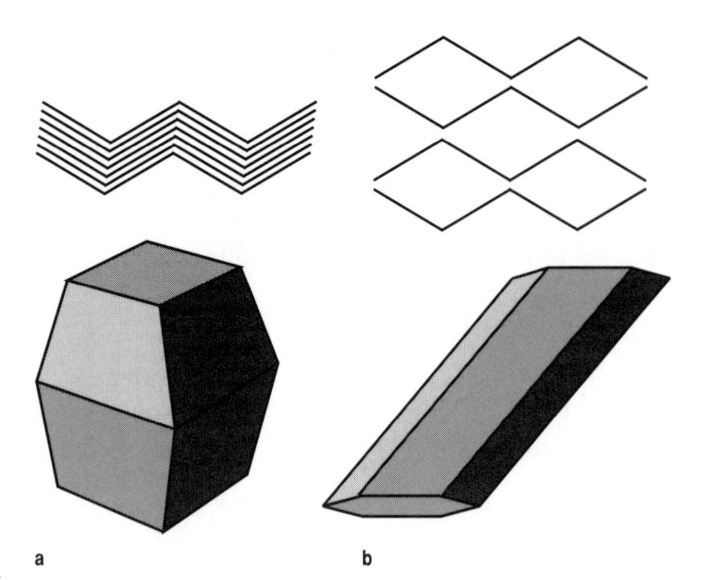

Figure 3.15.2 The packing of S_8 molecules in **a)** a rhombic crystal and **b)** a monoclinic crystal of sulphur

Uses of sulphur

The vast majority of sulphur is used to produce one of the most important industrial chemicals, sulphuric acid. Sulphur is also used to **vulcanise** rubber, a process which makes the rubber harder and increases its elasticity. Relatively small amounts are used in the manufacture of matches, fireworks, fungicides, as a sterilising agent and in medicines.

Allotropes of sulphur

Sulphur is one of the few non-metal elements which exist as allotropes. The main allotropes are called rhombic sulphur and monoclinic sulphur. Both of these solid forms of sulphur are made up of S_8 molecules.

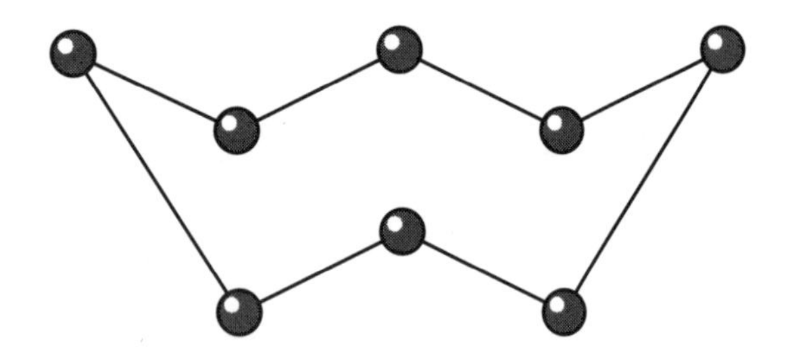

Figure 3.15.1 An S_8 molecule

The fact that there are two different allotropes of sulphur is due to the way in which these S_8 molecules pack together. In rhombic sulphur the molecules are packed more closely than in the monoclinic form.

Although sulphur is insoluble in water it will dissolve in an organic solvent such as xylene. If this solution is heated and allowed to cool down, then crystals of monoclinic sulphur are produced. When the temperature of the solution cools below 96 °C then rhombic sulphur crystals are produced. Rhombic sulphur is stable below 96 °C and monoclinic sulphur is stable above 96 °C. 96 °C is called the **transition temperature**.

When solid sulphur is heated, it melts at 112 °C and forms a runny liquid. At this point the S_8 molecules are moving freely around each other, the weak attractive forces between them having been overcome (see Figure 3.15.3).

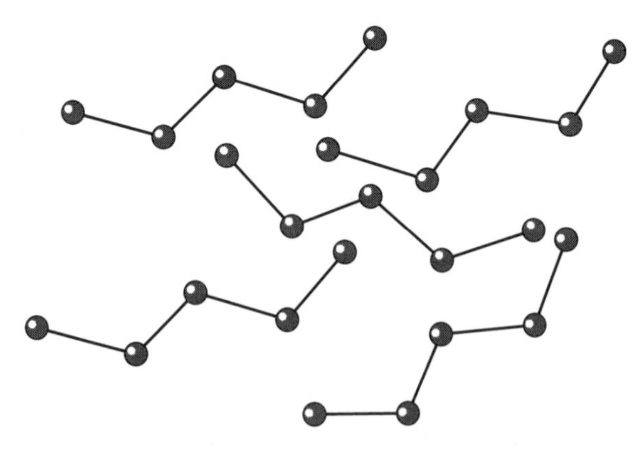

Figure 3.15.3 At the melting point S_8 rings move freely around one another

However if the sulphur is heated further, the liquid becomes thicker (more viscous). This is because the S_8 rings are broken by the extra added energy and they bond together forming long chains of sulphur atoms which become tangled, making the liquid viscous (see Figure 3.15.4).

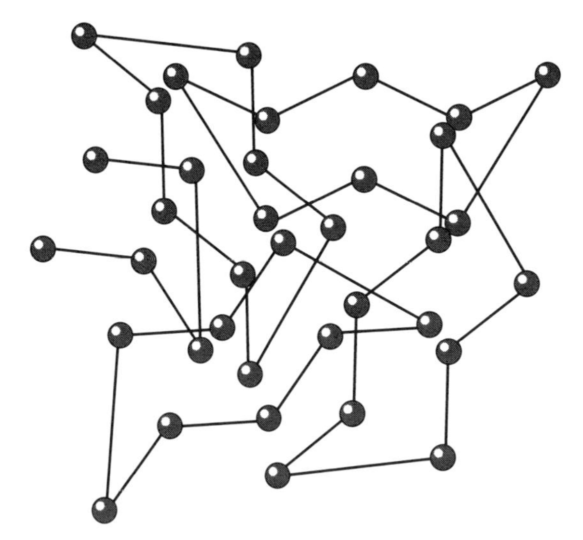

Figure 3.15.4 A viscous liquid is produced as chains of sulphur atoms are formed and get tangled together

Continued heating, to 444 °C, makes the liquid more mobile once again as the long chains are broken down into smaller ones which move around one another freely.

If this liquid is poured into a beaker of cold water, a substance called plastic sulphur is formed. This is an elastic, rubber-like substance. In plastic sulphur, the sulphur atoms remain bonded together in the form of chains – very similar to the chains of carbon atoms in plastics such as poly(ethene). After a few hours, however, the plastic sulphur loses its elasticity and once again becomes solid as the S_8 molecular rings reform.

Properties of sulphur

Physical properties

Sulphur:

- Is a yellow, brittle solid at room temperature.
- Does not conduct electricity.
- Is insoluble in water.

Chemical properties

Sulphur will react with both metals and non-metals.

- It reacts with magnesium to form magnesium sulphide.

$$\text{magnesium} + \text{sulphur} \longrightarrow \text{magnesium sulphide}$$
$$Mg(s) + S(s) \longrightarrow MgS(s)$$

- It reacts with oxygen to produce sulphur dioxide gas.

$$\text{sulphur} + \text{oxygen} \longrightarrow \text{sulphur dioxide}$$
$$S(s) + O_2(g) \longrightarrow SO_2(g)$$

Sulphur dioxide

Sulphur dioxide is a colourless gas which is produced when sulphur or substances containing sulphur, for example crude oil or natural gas, are burned in oxygen gas. It has a choking smell and is extremely poisonous. The gas dissolves in water to produce an acidic solution of sulphurous acid.

$$\text{sulphur dioxide} + \text{water} \rightleftharpoons \text{sulphurous acid}$$
$$SO_2(g) + H_2O(l) \rightleftharpoons H_2SO_3(aq)$$

It is one of the major pollutant gases and is the gas principally responsible for acid rain.

Sulphur dioxide is used as a fungicide and a preservative, as well as a bleaching agent in the paper industry.

Acid rain

Rainwater is naturally acidic, because it dissolves carbon dioxide gas from the atmosphere as it falls. Natural rainwater has a pH of about 5.7. But in recent years, especially in central Europe, the pH of rainwater has fallen to pH 3–4.8. This increase in acidity has led to extensive damage to forests, lakes and marine life.

The amount of sulphur dioxide in the atmosphere has increased dramatically over recent years. There was always some sulphur dioxide in the atmosphere, put there by natural processes such as volcano activity and rotting vegetation. Over Europe, however, around 80% of the sulphur dioxide in the atmosphere is formed from the combustion of fuels containing sulphur. After dissolving in rainwater to produce sulphurous acid, it further reacts with oxygen to produce sulphuric acid.

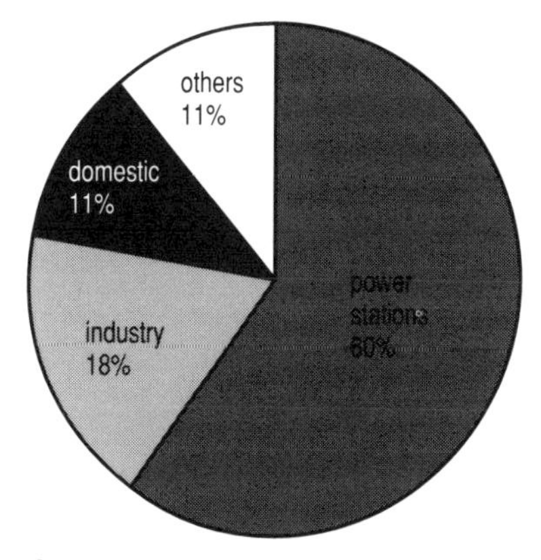

Figure 3.15.5 Sources of sulphur dioxide

Quick Questions

1. How could the amount of sulphur dioxide being produced by the above sources be reduced?
2. Devise an experiment which you could carry out in the school laboratory to determine the amount of sulphur in two different types of coal.
3. What is meant by the term allotrope?
4. 'Sulphur is a non-metallic element.' Discuss this statement giving physical and chemical reasons to support your answer.

3.16 Manufacture of sulphuric acid

The main use of sulphur is in the production of sulphuric acid. It is probably the most important industrial chemical produced in the world. Many millions of tonnes of sulphuric acid are produced in the UK each year. It is used mainly as the raw material for the production of many substances such as:

- Fertilisers
- Paints, pigments and dyes.
- Man-made fibres.
- Detergents.
- Chemicals.
- Car batteries.

The process by which sulphuric acid is produced is known as the **Contact process**.

The Contact process

Stage 1

Sulphur dioxide is first produced, primarily by burning sulphur with air.

$$\text{sulphur} + \text{oxygen} \longrightarrow \text{sulphur dioxide}$$
$$S(s) + O_2(g) \longrightarrow SO_2(g)$$

Stage 2

Any dust and impurities are removed from the sulphur dioxide, as well as any unreacted oxygen. These clean gases are heated to a temperature of approximately 450 °C and fed into a reaction vessel where they are passed over a catalyst of vanadium(V) oxide (V_2O_5). This catalyses the reaction to produce sulphur trioxide (sulphur(VI) oxide, SO_3).

$$\text{sulphur dioxide} + \text{oxygen} \rightleftharpoons \text{sulphur trioxide}$$
$$2SO_2(g) + O_2(g) \rightleftharpoons 2SO_3(g) \quad \Delta H = -197\ \text{kJ/mol}$$

This reaction is reversible and so the ideas of Le Chatelier (see Section 3.14) can be used to increase the proportion of sulphur trioxide in the equilibrium mixture.

The forward reaction is exothermic and so would be favoured by low temperatures. The temperature of 450 °C used is an optimum temperature which produces sufficient sulphur trioxide at an economic rate. Since the reaction from left to right is also accompanied by a decrease in the number of molecules of gas, it would be favoured by a high pressure. The process is actually run at just slightly above atmospheric pressure.

Under these conditions there is about 96% conversion into sulphur trioxide. The heat produced by this reaction is used to heat the incoming gases so saving money.

Stage 3

If this sulphur trioxide is directly added to water, sulphuric acid is produced. This reaction, however, is very violent and a thick mist is produced.

$$\text{sulphur trioxide} + \text{water} \longrightarrow \text{sulphuric acid}$$
$$SO_3(g) + H_2O(l) \longrightarrow H_2SO_4(l)$$

This acid mist is very difficult to deal with and so a different route to sulphuric acid is employed. Instead, the sulphur trioxide is dissolved in concentrated sulphuric acid (98%) to give a substance called oleum.

$$\text{sulphuric acid} + \text{sulphur trioxide} \longrightarrow \text{oleum}$$
$$H_2SO_4(aq) + SO_3(g) \longrightarrow H_2S_2O_7(l)$$

The oleum formed is then added to the correct amount of water to produce sulphuric acid of required concentration.

$$\text{oleum} + \text{water} \longrightarrow \text{sulphuric acid}$$
$$H_2S_2O_7(l) + H_2O(l) \longrightarrow 2H_2SO_4(l)$$

Properties of sulphuric acid

Dilute sulphuric acid

Dilute sulphuric acid is a typical strong acid. It will react with bases such as copper(II) oxide to produce salts, called sulphates, and water.

$$\text{copper(II) oxide} + \text{sulphuric acid} \longrightarrow \text{copper(II) sulphate} + \text{water}$$
$$CuO(s) + H_2SO_4(aq) \longrightarrow CuSO_4(aq) + H_2O(l)$$

It also reacts with carbonates to give salts, carbon dioxide and water, and with reactive metals to give a salt and hydrogen gas. The reaction between zinc and sulphuric acid is often used to prepare hydrogen gas in the laboratory.

$$\text{zinc} + \text{sulphuric acid} \longrightarrow \text{zinc sulphate} + \text{hydrogen}$$
$$Zn(s) + H_2SO_4(aq) \longrightarrow ZnSO_4(aq) + H_2(g)$$

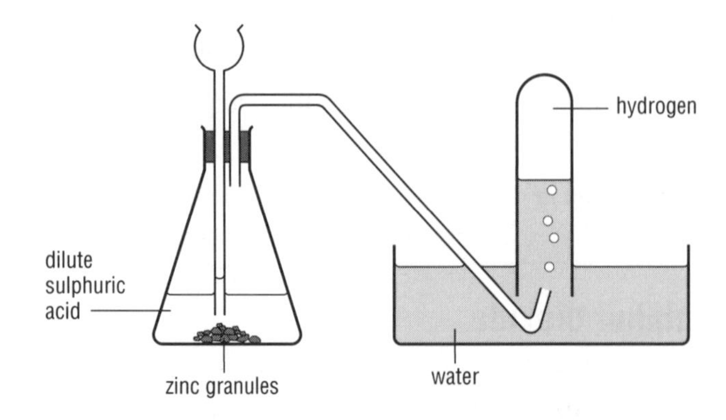

Figure 3.16.1 The laboratory preparation of hydrogen gas

Concentrated sulphuric acid

Concentrated sulphuric acid is a powerful dehydrating agent. That means it will take water from a variety of substances. One such substance is cane sugar, or sucrose.

$$\text{sucrose (sugar)} \xrightarrow{\text{conc } H_2SO_4} \text{carbon} + \text{water}$$
$$C_{12}H_{22}O_{11}(s) \longrightarrow 12C(s) + 11H_2O(l)$$

If a few drops of concentrated sulphuric acid are added to some blue hydrated copper(II) sulphate crystals they slowly turn white as the water of crystallisation is removed by the acid. Eventually it leaves only a white powder - anhydrous copper(II) sulphate.

$$\text{hydrated copper(II) sulphate} \xrightarrow{H_2SO_4} \text{anhydrous copper(II) sulphate} + \text{water}$$
$$\underset{\text{blue}}{CuSO_4.5H_2O(s)} \longrightarrow \underset{\text{white}}{CuSO_4(s)} + 5H_2O(l)$$

If concentrated sulphuric acid is added to a metal chloride such as sodium chloride, then hydrogen chloride gas is produced.

$$\text{sodium chloride} + \text{sulphuric acid} \longrightarrow \text{sodium hydrogensulphate} + \text{hydrogen chloride}$$
$$NaCl(s) + H_2SO_4(l) \longrightarrow NaHSO_4(s) + HCl(g)$$

If the hydrogen chloride is dissolved in water, dilute hydrochloric acid is produced.

Concentrated sulphuric acid should be treated very carefully because it will remove water from flesh! It is a very corrosive substance and should always be handled with care.

Because of its affinity for water the diluting of concentrated sulphuric acid must be carried out with care. The concentrated sulphuric acid should always be added to the water, not the other way around.

Sulphates

To test for a sulphate, add a few drops of dilute hydrochloric acid followed by a few drops of barium chloride. If a sulphate is present a white precipitate of barium sulphate is produced.

$$\text{barium ions} + \text{sulphate ions} \longrightarrow \text{barium sulphate}$$
$$Ba^{2+}(aq) + SO_4^{2-}(aq) \longrightarrow BaSO_4(s)$$

Many sulphates have very important uses.

Table 3.16.1 Uses of some metal sulphates

Salt	Formula	Use
Calcium sulphate	$CaSO_4.\frac{1}{2}H_2O$	'Plaster of Paris' used to set bones
Ammonium sulphate	$(NH_4)_2SO_4$	Fertiliser
Magnesium sulphate	$MgSO_4$	In medicine it is used as a laxative
Barium sulphate	$BaSO_4$	'Barium meal' used in diagnostic medical X-rays

Quick Questions

1 Produce a flow diagram to show the stages in the production of sulphuric acid by the Contact process. Include the essential conditions used in each stage.

2 Write word and balanced chemical equations for the reactions of dilute sulphuric acid with:
  a) Sodium hydroxide;
  b) Copper(II) carbonate;
  c) Magnesium;
  d) Zinc oxide.

3 Study the following reaction scheme:

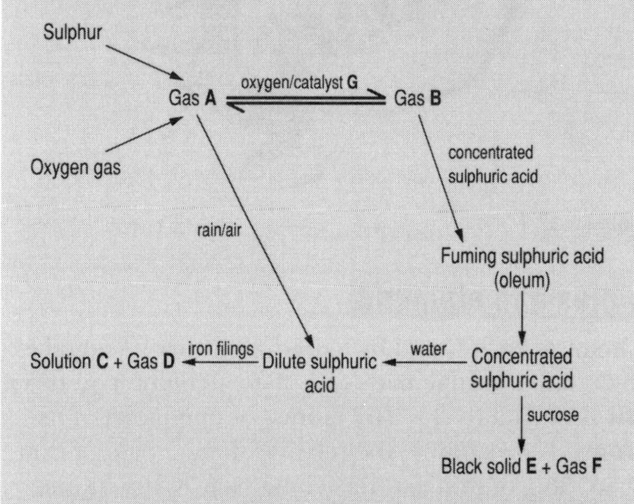

  a) Identify the substances A to G by giving their names and formulae.
  b) Write a balanced chemical equation for the formation of gas **B**.
  c) i) Describe a chemical test, and give the positive result of it, to identify gas **D**.
  ii) Describe a chemical test, and give the positive result of it, to identify gas **F**.
  d) How would you obtain solid **C** from the solution **C**?
  e) Which pathway shows the formation of acid rain?
  f) In which way is the concentrated sulphuric acid acting in its reaction with sucrose?
  g) Where does the oxygen gas come from to form gas **A**?

3.17 Quantitative chemistry

Chemists often need to know how much of a substance has been formed or used up during a chemical reaction. This is particularly important in the chemical industry where the substances you are reacting (the **reactants**) and the substances being produced (the **products**) are worth thousands of pounds. Waste costs money!

To solve this problem we need a way of counting atoms, ions or molecules. They are very tiny particles and it is impossible to measure out a dozen or even one hundred of them. Instead, chemists weigh out a very large number called a **mole** (often abbreviated to mol). A mole is 6×10^{23} atoms, ions or molecules. This number (6×10^{23}) is called **Avogadro's constant**. This constant was named after a famous Italian scientist called Amadeo Avogadro. So a mole of the element aluminium is 6×10^{23} atoms of aluminium and a mole of the element iron is 6×10^{23} atoms of iron.

Figure 3.17.1

Moles and elements

Chemists have found by experiment that if you take the relative atomic mass (A_r) of an element in grams it always contains 6×10^{23} atoms, or one mole, of its atoms. For example, the relative atomic mass of iron is 56. 56 g of iron and this contains 6×10^{23} atoms. Therefore 56 g of iron is a mole of iron atoms.

The mass of a substance present in any number of moles can be calculated using the relationship:

$$\text{mass (in grams)} = \text{number of moles} \times \text{mass of 1 mole of the element}$$

Examples

1 Calculate the mass of:
a) 2 moles;
b) 0.25 moles of iron
(A_r: Fe = 56).

a) mass of 2 moles of iron
= number of moles × relative atomic mass (A_r)
= 2 × 56 = 112 g

b) mass of 0.25 moles of iron
= number of moles × relative atomic mass (A_r)
= 0.25 × 56 = 14 g

If we know the mass of the element then it is possible to calculate the number of moles of that element using:

$$\text{number of moles} = \frac{\text{mass of the element}}{\text{mass of 1 mole of that element}}$$

2 Calculate the number of moles of aluminium present in:
a) 108 g; and
b) 13.5 g of the element
(A_r: Al = 27).

a) number of moles of aluminium

$$= \frac{\text{mass of aluminium}}{\text{mass of 1 mole of aluminium}} = \frac{108}{27} = 4$$

b) number of moles of aluminium

$$= \frac{\text{mass of aluminium}}{\text{mass of 1 mole of aluminium}} = \frac{13.5}{27} = 0.5$$

Moles and compounds

The mole can also be used with compounds. We cannot discuss the atomic mass of a molecule or of a compound because more than one type of atom is involved. Instead we have to discuss the **relative formula mass** (RFM), often called the relative molecular mass (M_r) when applied to covalent substances. This is the sum of the relative atomic masses of all those elements shown in the formula of the substance.

Examples

1 What is the relative formula mass of water (H_2O) molecules? (A_r: H = 1, O = 16).

From the formula of water, H_2O, you will see that one mole of water molecules contains 2 moles of hydrogen (H) atoms and 1 mole of oxygen (O) atoms. The relative formula mass of water molecules is therefore:

$$(2 \times 1) + (1 \times 16) = 18$$

The mass of one mole of a compound is called its **molar mass**. If you write the molar mass of a compound without any units then it is the relative formula mass or the relative molecular mass. So the relative molecular mass of water is 18.

2 What is:

a) the RFM;

b) the mass of one mole of ethanol, C_2H_5OH? (A_r: C = 12; H = 1; O = 16.)

a) One mole of C_2H_5OH contains:
2 moles of carbon atoms
6 moles of hydrogen atoms
1 mole of oxygen atoms
Therefore,
the RFM of ethanol
$= (2 \times 12) + (6 \times 1) + (1 \times 16)$
$= 46$

b) The mass of one mole of ethanol is 46 g.

The mass of a compound found in any number of moles can be calculated using the relationship:

$$\text{mass of compound} = \text{number of moles of compound} \times \text{mass of 1 mole of compound}$$

3 Calculate the mass of

a) 3 moles;

b) 0.2 moles of carbon dioxide gas, CO_2. (Ar: C = 12; O = 16.)

a) One mole of CO_2 contains:
1 mole of carbon atoms
2 moles of oxygen atoms

Therefore,
the mass of 1 mole of CO_2
$= (1 \times 12) + (2 \times 16) = 44$ g

Mass of 3 moles of CO_2
= no. of moles × mass of 1 mole of CO_2
$= 3 \times 44 = 132$ g

b) Mass of 0.2 moles of CO_2
= no. of moles × mass of 1 mole of CO_2
$= 0.2 \times 44 = 8.8$ g

If we know the mass of the compound then it is possible to calculate the number of moles of the compound using the relationship:

$$\text{no. of moles of compound} = \frac{\text{mass of compound}}{\text{mass of 1 mole of compound}}$$

4 Calculate the number of moles of magnesium oxide, MgO, in 80 g of the compound. (A_r: Mg = 24; O = 16.)

One mole of MgO contains:
1 mole of magnesium atoms
1 mole of oxygen atoms

Therefore,
the mass of one mole of MgO
$= (1 \times 24) + (1 \times 16) = 40$ g

Number of moles of MgO in 80 g

$$= \frac{\text{mass of MgO}}{\text{mass of 1 mole of MgO}} = \frac{80}{40} = 2$$

Quick Questions

1 Calculate the number of moles in:
a) 2 g of neon atoms;
b) 4 g of magnesium atoms;
c) 24 g of carbon atoms.

2 Calculate the mass of:
a) 0.1 moles of oxygen molecules;
b) 5 moles of sulphur atoms;
c) 0.25 moles of sodium atoms.

3 Calculate the number of moles in:
a) 9.8 g of sulphuric acid (H_2SO_4);
b) 40 g of sodium hydroxide (NaOH);
c) 720 g of iron(II) oxide (FeO).

4 Calculate the mass of:
a) 2 moles of zinc oxide (ZnO);
b) 0.25 moles of hydrogen sulphide (H_2S);
c) 0.35 moles of copper(II) sulphate ($CuSO_4$).

Moles and gases

Many substances exist as gases. If we want to find the number of moles of a gas we can do this by measuring the volume rather than the mass.

Chemists have shown by experiment that: one mole of any gas occupies a volume of approximately 24 dm^3 (24 l) at room temperature and pressure (rtp).

Therefore, it is relatively easy to convert volumes of gases into moles and moles of gases into volumes using the following relationship:

$$\text{no. of moles of a gas} = \frac{\text{vol. of the gas (in dm}^3\text{ at rtp)}}{24\ \text{dm}^3}$$

or,

$$\text{vol. of a gas (in dm}^3\text{ at rtp)} = \text{no. of moles of gas} \times 24\ \text{dm}^3$$

Example

Calculate the moles of ammonia gas, NH_3, in a volume of 72 dm^3 of the gas measured at rtp.

$$\text{no. of moles of ammonia} = \frac{\text{volume of ammonia in dm}^3}{24\ \text{dm}^3}$$

$$= \frac{72}{24} = 3$$

The volume occupied by one mole of any gas must contain 6×10^{23} molecules. Therefore, it follows that equal volumes of all gases measured at the same temperature and pressure must contain the same number of molecules. This idea was first put forward by Amadeo Avogadro and is called **Avogadro's law**.

Moles and solutions

Chemists often need to know the concentration of a solution. Concentration is measured in moles per dm^3 (mol/dm^3). When one mole of a substance is dissolved in water and the solution made up to 1 dm^3 (1000 cm^3) a 1 molar (1 M or 1 mol/dm^3) solution is produced. Chemists do not always need to make up such large volumes of solution. A simple method of calculating the concentration is by using the relationship:

$$\text{concentration} = \frac{\text{number of moles}}{\text{volume (in dm}^3\text{)}}$$

Example
Calculate the concentration (in mol/dm^3) of a solution of sodium hydroxide, NaOH, which was made by dissolving 10 g of solid sodium hydroxide in water and making up to 250 cm^3. (A_r: Na = 23; O = 16; H = 1.)

1 mole of NaOH contains:
1 mole of sodium
1 mole of oxygen
1 mole of hydrogen

Therefore,
mass of one mole of NaOH

$= (1 \times 23) + (1 \times 16) + (1 \times 1) = 40$ g

number of moles of NaOH in 10 g

$$= \frac{\text{mass of NaOH}}{\text{mass of 1 mole of NaOH}}$$

$$= \frac{10}{40} = 0.25$$

concentration of the NaOH solution

$$= \frac{\text{number of moles of NaOH}}{\text{volume of solution (in dm}^3\text{)}}$$

$$[250 \text{ cm}^3 = \frac{250}{1000} \text{ dm}^3 = 0.25 \text{ dm}^3]$$

$$= \frac{0.25}{0.25} = 1 \text{ mol dm}^3 \text{ (or 1 M)}$$

Sometimes chemists need to know the mass of a substance that has to be dissolved to prepare a known volume of solution of a given concentration. A simple method of calculating the number of moles and hence the mass of substance needed is by using the relationship:

$$\text{number of moles} = \text{concentration (in mol/dm}^3\text{)} \times \text{volume of solution (in dm}^3\text{)}$$

Example
Calculate the mass of potassium hydroxide, KOH, which needs to be used to prepare 500 cm^3 of a 2 mol/dm^3 (2 M) solution in water. (A_r: K = 39; 0 = 16; H = 1.)

number of moles of KOH
= concentration of solution × volume of solution (dm^3)

$$= 2 \times \frac{500}{1000} = 1$$

1 mole of KOH contains:
1 mole of potassium
1 mole of oxygen
1 mole of hydrogen

Therefore,
mass of 1 mole of KOH
$= (1 \times 39) + (1 \times 16) + (1 \times 1) = 56$ g

Therefore, the mass of KOH in one mole
= number of moles × mass of 1 mole
$= 1 \times 56 = 56$ g

Titration

A **titration** is a technique used by chemists to find the concentration of a substance. For example, if we wanted to know the concentration of a solution of sodium hydroxide the titration of hydrochloric acid with sodium hydroxide would be carried out in the following way.

25 cm^3 of sodium hydroxide solution is pipetted into a conical flask, to which a few drops of phenolphthalein indicator has been added. Phenolpthalein is pink in alkali and colourless in acid.

0.1 M hydrochloric acid is placed in the burette using a filter funnel until it is filled up exactly to the zero mark.

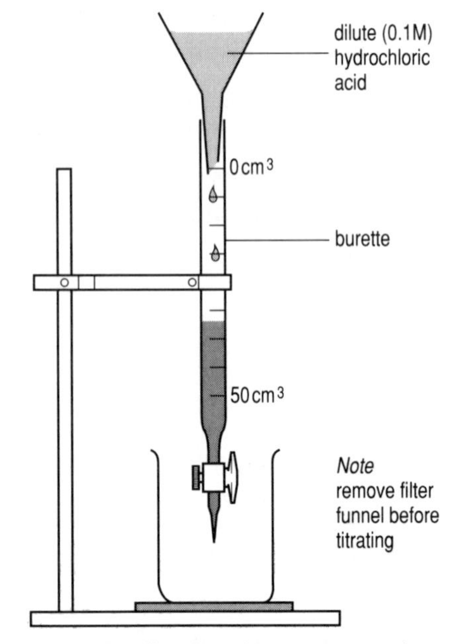

Figure 3.17.2 The filter funnel is now removed

The hydrochloric acid is added to the sodium hydroxide solution in small quantities – usually no more than 0.5 cm^3 at a time. The contents of the flask must be swirled after each addition of acid.

The acid is added until the alkali has been neutralised completely. This is shown by the pink colour of the indicator just disappearing.

The final reading on the burette at the end-point is recorded and further titrations carried out until consistent results are obtained (within 0.1 cm^3 of each other). Below are shown some sample data.

Figure 3.17.3

Volume of sodium hydroxide solution $= 25\ cm^3$.

Average volume of 0.1 M hydrochloric acid added $= 22.0\ cm^3$.

The neutralisation reaction which has taken place is:

hydrochloric acid	+	sodium hydroxide	⟶	sodium chloride	+	water
$HCl(aq)$	+	$NaOH(aq)$	⟶	$NaCl(aq)$	+	$H_2O(l)$

From this equation it can be seen that 1 mole of hydrochloric acid neutralises 1 mole of sodium hydroxide. Now you can work out the number of moles of the acid using the formula given earlier.

$$\text{moles of acid} = \text{concentration} \times \frac{\text{volume}}{1000}$$

$$= 0.1 \times \frac{22.0}{1000} = 0.0022$$

$$\text{number of moles of hydrochloric acid} = \text{number of moles of sodium hydroxide}$$

Therefore, the number of moles of sodium hydroxide = 0.0022.

0.0022 moles of sodium hydroxide is present in 25 cm^3 of solution.

Therefore in 1 dm^3 of sodium hydroxide solution we have

$$\frac{0.0022 \times 1000}{25} = 0.088 \text{ moles}$$

The concentration of sodium hydroxide solution $= 0.088\ mol/dm^3$. Therefore it is a 0.088 M solution of sodium hydroxide.

Quick Questions

Use the values of A_r which follow to answer the questions below:

C = 12; Ne = 20; Mg = 24; O = 16; S = 32; Na = 23; H = 1; Fe = 56; Cu = 63.5; N = 14; Zn = 65; K = 39.

One mole of any gas at room temperature and pressure occupies 24 dm^3.

1 Calculate the number of moles at room temperature and pressure in:
- **a)** 2 dm^3 of carbon dioxide;
- **b)** 20 cm^3 of carbon monoxide.

2 Calculate the volume of:
- **a)** 0.3 moles of hydrogen chloride;
- **b)** 34 g of ammonia (NH_3).

3 Calculate the concentration of solutions containing:
- **a)** 0.2 moles of sodium hydroxide dissolved in water and made up to 100 cm^3;
- **b)** 9.8 g of sulphuric acid (H_2SO_4) dissolved in water and made up to 500 cm^3.

4 Calculate the mass of copper(II) sulphate ($CuSO_4$) which needs to be used to prepare 500 cm^3 of a 0.1 M solution.

5 28 cm^3 of 0.200 M hydrochloric acid just neutralised 25 cm^3 of a potassium hydroxide solution. What is the concentration of this potassium hydroxide solution?

Calculating formulae

If we have one mole of a compound then the formula shows the number of moles of each element in that compound. For example, the formula for lead(II) bromide is $PbBr_2$. This means that 1 mole of lead(II) bromide contains 1 mole of lead and 2 moles of bromine. If we do not know the formula of a compound we can find the masses of the elements present experimentally and these masses can be used to work out the formula of that compound.

Finding the formula of magnesium oxide

When magnesium ribbon is heated strongly it burns very brightly to form the white powder called magnesium oxide.

magnesium + oxygen ⟶ magnesium oxide

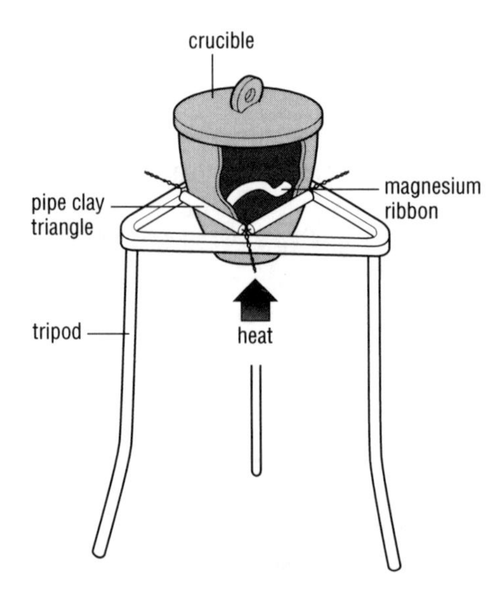

Figure 3.17.4 Apparatus used to determine magnesium oxide's formula

The following data shows how the mass of magnesium increases when it is heated, as it forms the oxide.

Mass of crucible	= 14.63 g
Mass of crucible and magnesium	= 14.87 g
Mass of crucible and magnesium oxide	= 15.03 g
Mass of magnesium used	= 0.24 g
Mass of oxygen which has reacted with the magnesium	= 0.16 g

From this data we can calculate the number of moles of each of the reacting elements. (A_r: Mg = 24; O = 16.)

	Mg	O
masses reacting/g	0.24	0.16
number of moles	$\frac{0.24}{24}$	$\frac{0.16}{16}$
	= 0.01	= 0.01
ratio of moles	1	1
formula	MgO	

This formula is called the **empirical formula** of the compound. It shows the simplest ratio of the atoms present.

Example

In another experiment an unknown organic compound was found to contain 0.12 g of carbon and 0.02 g of hydrogen. Calculate the empirical formula of the compound. (A_r: C = 12; H = 1.)

	C	H
masses/g	0.12	0.02
number of moles	$\frac{0.12}{12}$	$\frac{0.02}{1}$
	= 0.01	= 0.02
ratio of moles	1	2
empirical formula	CH_2	

However, from our knowledge of bonding, a molecule of this formula cannot exist, but molecules of the following formulae do exist; C_2H_4, C_3H_6, and C_4H_8. All of these formulae show the same ratio of carbon atoms to hydrogen atoms - CH_2, as our example. To find out which of these formulae is the actual formula for the unknown organic compound we need to know the mass of one mole of the compound.

Using a mass spectrometer the relative molecular mass (M_r) of this organic compound was found to be 56. We need to find out the number of empirical formulae units present.

M_r of the empirical formula unit - CH_2
$= (1 \times 12) + (2 \times 1) = 14$

Empirical formula units present

$$= \frac{M_r \text{ of the compound}}{M_r \text{ of empirical formula unit}} = \frac{56}{14} = 4$$

Therefore, the actual formula, the **molecular formula**, of the unknown organic compound is $4 \times CH_2 = C_4H_8$.

Sometimes the composition of a compound is given as a percentage by mass of the elements present. In cases such as this assume that 100 g of the compound is present so that the percent of each element present will numerically equal the mass in grams.

Moles and chemical equations

When we write a balanced chemical equation we are indicating the numbers of moles of reactants and products involved in the chemical reaction. Consider the reaction between magnesium and oxygen.

magnesium	+	oxygen	$\longrightarrow$	magnesium oxide
$2Mg(s)$	+	$O_2(g)$	$\longrightarrow$	$2MgO(s)$

This shows that 2 moles of magnesium react with 1 mole of oxygen to give 2 moles of magnesium oxide.

Using the ideas of moles and masses we can use this information to calculate the quantities of the different chemicals involved.

$2Mg(s)$	+	$O_2(g)$	$\longrightarrow$	$2MgO(s)$
2 moles		1 mole		2 moles
2×24		$1 \times (16 \times 2)$		$2 \times (24 + 16)$
= 48 g		= 32 g		= 80 g

You will notice that the total mass of reactants is equal to the total mass of product. This is true for any chemical reaction and it is known as the **law of conservation of mass** for a chemical reaction. Chemists can use this idea to calculate masses of products formed and reactants used in chemical processes before they are carried out.

Example

Lime (calcium oxide, CaO) is used in the manufacture of mortar. It is manufactured in large quantities by Tilcon at the Swindon Quarry in North Yorkshire by heating limestone (calcium carbonate, $CaCO_3$). The equation for the process is:

$CaCO_3(s)$	$\longrightarrow$	$CaO(s)$	+	$CO_2(g)$
1 mole		1 mole		1 mole
$(40 + 12 + (3 \times 16))$		$40 + 16$		$12 + (2 \times 16)$
100 g		56 g		44 g

Calculate the amount of lime produced when 10 tonnes of limestone are heated. (A_r: Ca = 40; C = 12; O = 16.)

1 tonne = 1000 kg

1 kg = 1000 g

From this relationship between grams and tonnes we can replace the masses in grams by masses in tonnes.

$CaCO_3(s)$	$\longrightarrow$	$CaO(s)$	+	$CO_2(g)$
100 t		56 t		44 t
10 t		5.6 t		4.4 t

The equation now shows that 100 t of limestone will produce 56 t of lime. Therefore, 10 t of limestone will produce 5.6 t of lime.

Many chemical processes involve gases. The volume of a gas is measured more easily than its mass. The next example shows how chemists work out the volumes of gaseous reactants and products needed using ideas of moles.

Some rockets use hydrogen gas as a fuel. When hydrogen burns in oxygen it forms steam.

Calculate the volumes of **a)** water $H_2O(g)$ produced; **b)** $O_2(g)$ used if 960 dm^3 of hydrogen gas, $H_2(g)$, were burned in oxygen. (A_r: H = 1; O = 16.) Assume 1 mole of any gas occupies a volume of 24 dm^3 at rtp.

$2H_2(g)$	+	$O_2(g)$	$\longrightarrow$	$2H_2O(g)$
2 moles		1 mole		2 moles
2×24		1×24		2×24
= 48 dm^3		= 24 dm^3		= 48 dm^3

Therefore

(×2)	96 dm^3	48 dm^3	96 dm^3
(×10)	960 dm^3	480 dm^3	960 dm^3

When 960 dm^3 of hydrogen are burned in oxygen:

a) 480 dm^3 of oxygen are required;

b) 960 dm^3 of $H_2O(g)$ are produced.

Examples using a solution

Chemists usually carry out chemical reactions using solutions. If they know the concentration of the solution(s) they are using then they can find out the quantities reacting.

Calculate the volume of 1 M H_2SO_4 required to react completely with 6 g of magnesium. (A_r: Mg = 24.)

$$\text{Number of moles of Mg} = \frac{6}{24} = 0.25$$

$Mg(s)$	+	$H_2SO_4(aq)$	$\longrightarrow$	$MgSO_4(aq)$	+	$H_2(g)$
1 mole		1 mole		1 mole		1 mole
0.25 mol		0.25 mol		0.25 mol		0.25 mol

We can see that 0.25 mol of $H_2SO_4(aq)$ are required. Using:

$$\text{volume of } H_2SO_4(aq)\ (dm^3) = \frac{\text{moles of } H_2SO_4}{\text{conc. of } H_2SO_4\ (mol/dm^3)}$$

$$= \frac{0.25}{1}$$

$$= 0.25\ dm^3 \text{ or } 250\ cm^3.$$

Quick Questions

Use the following A_r values to answer the questions which follow: O = 16; S = 32; Ca = 40.

1. Calculate the mass of sulphur dioxide produced by burning 16 g of sulphur in an excess of oxygen.
2. Determine the empirical formula of an oxide of calcium formed when 0.4 g of calcium reacts with 0.16 g of oxygen.

Section Three: Examination Questions

1. **(a)** Limestone, basalt, slate, marble, sandstone and granite are all examples of different rocks.
From the above list of rocks choose and give the name of:
(i) a sedimentary rock;
(ii) a metamorphic rock;
(iii) an igneous rock. [3]

(b) Which type of rock, sedimentary, metamorphic or igneous, is formed when molten rock (magma) is forced up from inside the Earth? [1]

(c) Describe how sedimentary rock is formed over long periods of time. [3]

(d) Rocks are widely used in the construction industry. Complete the table below by choosing from the uses listed in the box.

building; cement; concrete; road making.

Each use can only be selected once. One use has been completed for you.

Rock	*Use*
slate	building
granite	
limestone	
sand	

[3]
(WJEC, Higher, 1998 Specimen)

2. Calcium carbonate, $CaCO_3$, is found in limestone rock.
(a) Name two other rocks which are mostly calcium carbonate. [2]
(b) Limestone can be changed into other useful substances as shown below.

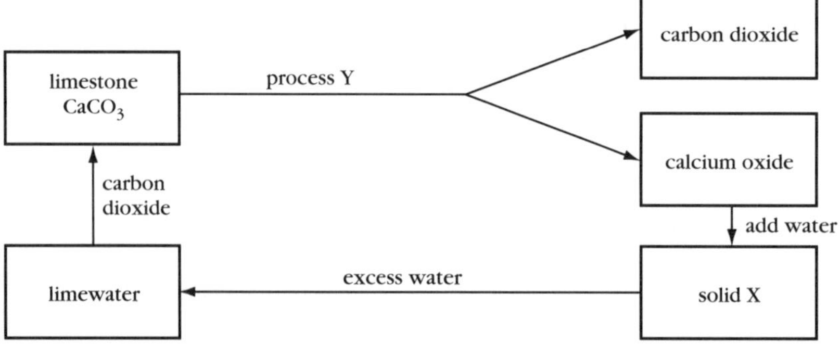

(i) Name process Y. [1]
(ii) Name solid X. [1]
(iii) [A] What is seen when carbon dioxide is bubbled through limewater? [1]
[B] What use is this reaction? [1]

(c) State **two** uses of carbon dioxide [2]

(SEG, Foundation, 1998 Specimen)

3. **(a)** Crude oil is used in the production of many useful substances.

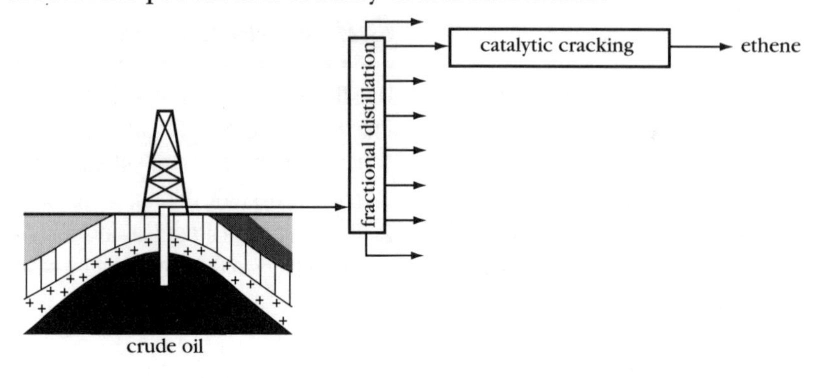

(i) Explain how crude oil was formed. [3]
(ii) How is crude oil extracted from the ground? [1]
(iii) Explain how fractional distillation can be used to separate compounds in crude oil. [2]
(iv) Explain how catalytic cracking can be used to make ethene, C_2H_4. [2]

(b) Ethene can be made into poly(ethene).
(i) Draw the structural formula of ethene. [1]
(ii) Name the process by which poly(ethene) is made. [1]
(iii) Explain what happens to the ethene molecules during the process to make poly(ethene). [2]

(c) Methane, CH_4, is often found in the same underground areas as crude oil. Relative atomic masses are: C 12; H 1.
(i) Calculate the relative formula mass of methane, CH_4. [1]
(ii) Calculate the percentage of carbon in a molecule of methane. [2]

(SEG Double Award, Higher, 1998 Specimen)

4. **(a)** Carbohydrates are natural polymers. Carbohydrates can be fermented under anaerobic conditions to make ethanol.
(i) What is meant by the term *polymer*? [2]
(ii) What is meant by the term *anaerobic*? [2]
(iii) State **one** use of ethanol. [1]
(iv) Fill in the gaps in the flow chart. [2]

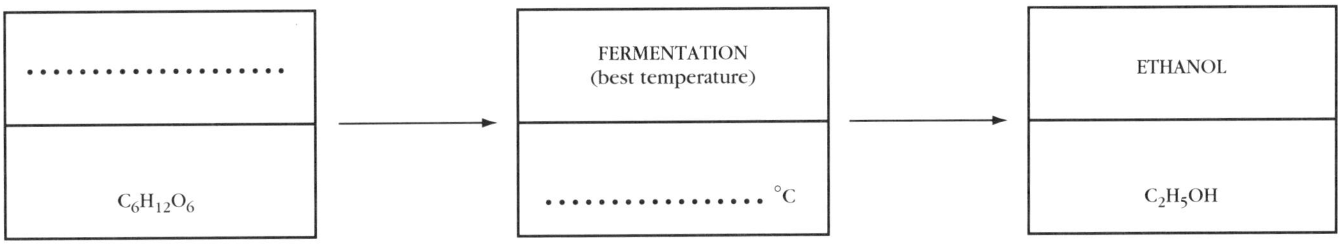

....	→	FERMENTATION (best temperature)	→	ETHANOL
$C_6H_{12}O_6$		.... °C		C_2H_5OH

(v) What must be added to a carbohydrate solution for fermentation to occur? [1]

(b) Wine contains ethanol. A glass of wine was found to have a sour taste. This was due to the presence of substance Z, formula CH_3COOH.
(i) What gas in the air caused the wine to change? [1]
(ii) Why did substance Z make the wine taste sour? [1]
(iii) Name substance Z. [1]

(c) Ethanol vapour was passed over hot aluminium oxide in the apparatus shown to produce the gas ethene.

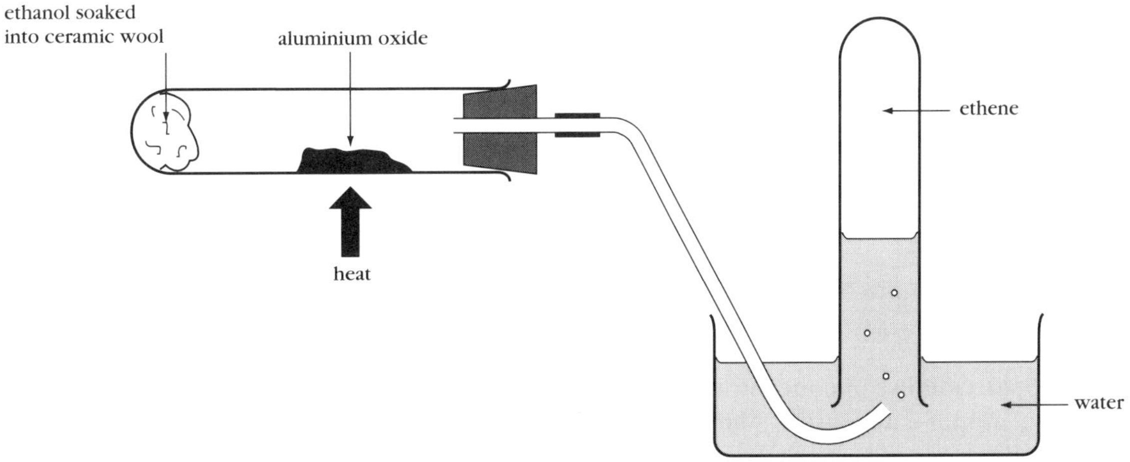

ethanol soaked into ceramic wool
aluminium oxide
heat
ethene
water

(i) Why is the reaction called a *dehydration*? [1]
(ii) The aluminium oxide acts as a dehydrating agent. Name one other dehydrating agent. [1]
(iii) Write a word equation for the formation of ethene from ethanol. [1]
(iv) What is seen when ethene is passed through bromine water? [1]

(SEG, Foundation, 1998 Specimen)

5. The figure shows a series of experiments set up to investigate the rusting of iron.

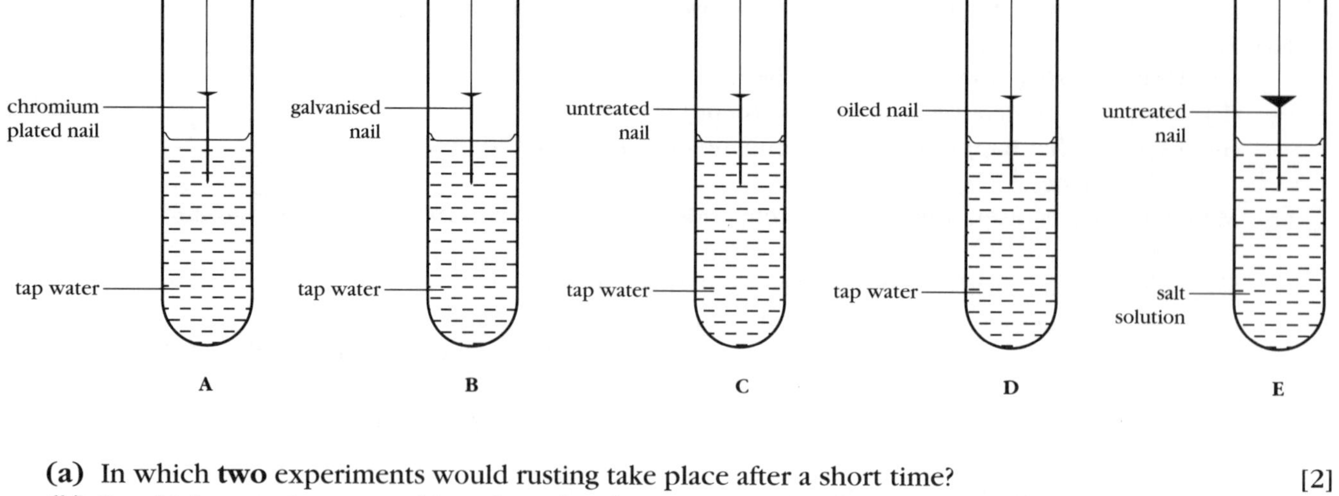

(a) In which **two** experiments would rusting take place after a short time? [2]
(b) In which experiment would rusting take place most quickly? [1]
(c) In which experiment had the nails received a treatment which could be used on:
bicycle handlebars? [1]
metal dustbins? [1]
(d) What process could be used to obtain salt from the solution in E? [1]

(MEG, Foundation, 1998 Specimen)

6. When manufacturing sulphuric acid, sulphur dioxide is firstly made into sulphur trioxide.

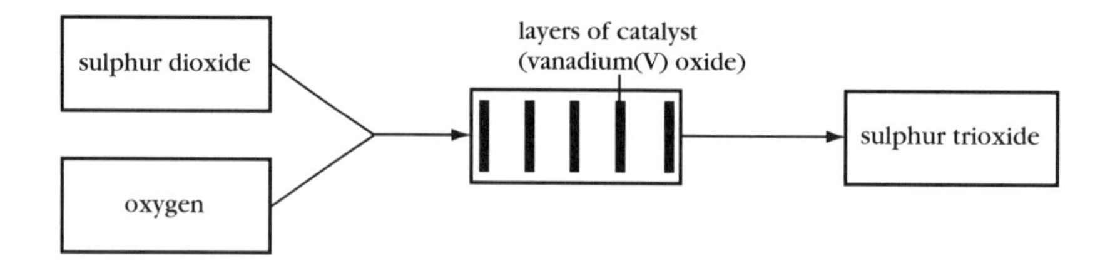

(a) (i) Give **two** reasons why a catalyst is used in this reaction. [2]
(ii) Write the word equation for making sulphur dioxide into sulphur trioxide. [1]
(b) (i) Which **one** of these warning signs would be found on a bottle of concentrated sulphuric acid? [1]

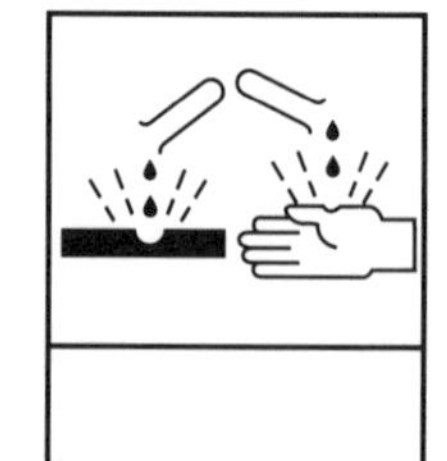

(ii) State **one** use of concentrated sulphuric acid. [1]
(c) When copper carbonate is added to cold dilute sulphuric acid it fizzes quickly and a blue solution is formed.
(i) What type of chemical reaction occurs when copper carbonate is added to dilute sulphuric acid? [1]
(ii) Why is the sulphuric acid not heated? [1]
(iii) Name the blue solution formed. [1]

(SEG, Foundation, 1998 Specimen)

7. (a) An atom has 11 protons, 11 electrons and 12 neutrons.

(i) Which **two** particles contribute almost entirely to the mass of the atom? [2]

(ii) What is the mass number of the atom? [1]

(b) The relative atomic mass of magnesium is 24. What standard is used in determining relative atomic masses? [2]

(c) Calculate the relative formula masses for the following compounds. (Relative atomic masses: H = 1, C = 12, O = 16, Na = 23, S = 32, Cl = 35.5, Ca = 40, Cu = 64.)

(i) NaCl [1]

(ii) $CaCl_2$ [1]

(iii) C_6H_6 [1]

(iv) $CuSO_4.5H_2O$ [3]

(d) A compound MBr_4 has a relative formula mass of 439.

(i) What is the relative atomic mass of M?
(Relative atomic mass: Br = 80) [3]

(ii) Which element does M represent? (You should use the Chemistry Data Tables at the end of this book to help you to answer this question.) [1]

(NICCEA, Foundation, 1998 Specimen)

8. Until fairly recently scientists thought that the continents had always been in the same positions.

Diagram 1 shows where scientists now think the continents were 200 million years ago. Diagram 2 shows the present positions of the continents.

Diagram 1

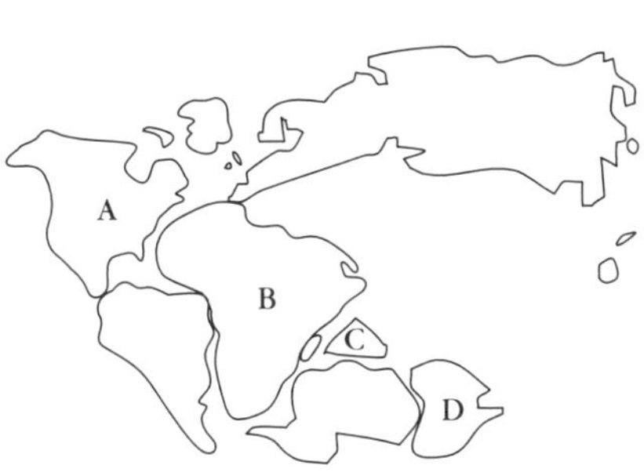

A B C D

Diagram 2

(a) Four land masses have been labelled A, B, C and D on Diagram 1.

Use the same letters A, B, C and D, **to label on Diagram 2** the present day positions of the four land masses. [2]

(b) The theory of crustal movements states that the continents have moved apart over the last 200 million years. Give **two** pieces of evidence for the movement. [2]

(c) The crust of the Earth includes several tectonic plates.

What causes tectonic plates to move? [3]

(d) The tectonic plate labelled D in diagram 1 eventually collided with an oceanic tectonic plate.

Describe what happens to tectonic plates when they collide. [4]

(NEAB, Foundation, 1998 Specimen)

9. On a hill walk a student found a rock in a stream. The student thought that the rock might be limestone.

(a) Use the Chemistry Data Tables at the end of this book to help you suggest a test for limestone. [2]

(b) Further tests back at school showed that the rock contained a compound with the formula $CaCO_3$.

Use the Chemistry Data Tables at the end of this book to help you answer the following questions.

(i) Name the **three** elements that are in $CaCO_3$. [3]

(ii) Calculate the relative formula mass (M_r) of $CaCO_3$. [2]

(c) The rock was limestone. Limestone is a sedimentary rock which contains the shelly remains of living organisms.

Describe how these shelly remains were changed into limestone. [3]

(d) Marble is a metamorphic rock formed from limestone.

Describe how marble is formed from limestone. [3]

(NEAB, Higher, 1998 Specimen)

10. Crude oil is a mixture of many compounds. Most of these compounds are hydrocarbons. The structure of one of these compounds is shown in the diagram.

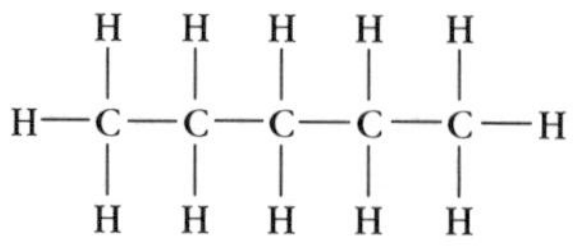

```
    H   H   H   H   H
    |   |   |   |   |
H — C — C — C — C — C — H
    |   |   |   |   |
    H   H   H   H   H
```

(a) What is a hydrocarbon? [1]

(b) What is the chemical formula of the structure shown in the diagram? [1]

(c) Crude oil consists of a large number of different compounds. Fractional distillation and cracking can be used to produce useful compounds from crude oil.

Describe, in as much detail as you can, how these two processes produce alkanes that are useful as fuels. [5]

(d) Ethene is an unsaturated hydrocarbon. What is meant by the term 'unsaturated'? [1]

(e) Complete the following equation to show how three ethene molecules join together to form part of a poly(ethene) molecule. [2]

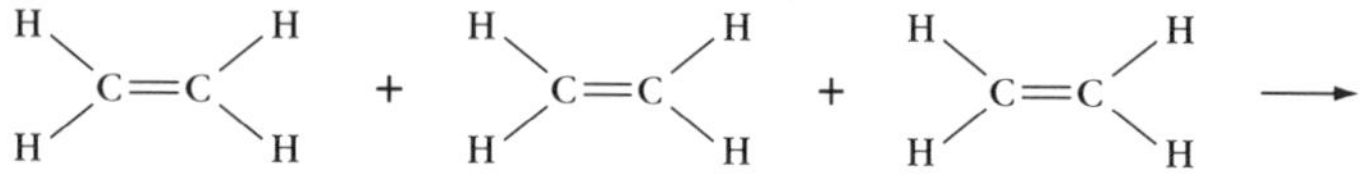

```
H       H     H       H     H       H
 \     /       \     /       \     /
  C = C    +    C = C    +    C = C    ——→
 /     \       /     \       /     \
H       H     H       H     H       H
```

(f) (i) Suggest one property of poly(ethene) which makes it suitable as a material for food containers. [1]

(ii) Thermosetting plastics cannot be re-moulded after they have been heated and allowed to cool. Explain why. [2]

(NEAB, Higher, 1998 Specimen)

11. (a) An important process used in the petrochemical industry is called *'cracking'*. The products are used as fuels or chemical feedstocks (petrochemicals).

An example of this process is given in the following equation, for the cracking of decane to give octane and ethene, which are described as being saturated and unsaturated hydrocarbons respectively.

$$C_{10}H_{22}(l) \longrightarrow C_8H_{18}(l) + C_2H_4(g)$$

(i) What do the symbols (l) and (g) mean in the above equation? [1]

(ii) What is the difference between a saturated and an unsaturated hydrocarbon? [1]

(b) Ethene undergoes addition polymerisation to form poly(ethene) (polythene), which is a plastic.

(i) Write down what is meant by addition polymerisation. [2]

(ii) Give a symbol equation for the reaction occurring, in the addition polymerisation of ethene. [2]

(iii) Give **one** use of polyethene (polythene). [1]

(c) The table below gives some information about four different plastics, **A**, **B**, **C** and **D**.

	Effect of heat	*Flexibility*	*Hardness*	*Colour*
A	melts	brittle	soft	white
B	melts	very flexible	soft	white
C	melts	brittle	hard	transparent
D	stable	brittle	hard	various

Write down which plastic you would choose for making (i) a plastic shopping bag and (ii) a fish bowl, giving **one** reason in each case. [4]

(WJEC, Higher, 1998 Specimen)

12. Ammonia, NH_3, is made by reacting together hydrogen and nitrogen in the presence of iron. This reaction is called the Haber process.

(a) How does the presence of iron help the process? [1]

(b) The table shows how much ammonia is produced using different conditions.

	percentage yield of ammonia at these temperatures		
pressure/atm	100 °C	300 °C	500 °C
25	91.7	27.4	2.9
50	94.5	39.5	5.6
100	96.7	52.5	10.6
200	98.4	66.7	18.3
400	99.4	79.7	31.9

From the values in the table, what happens to the yield of ammonia as:

(i) the temperature is increased? [1]

(ii) the pressure is increased? [1]

(c) Using ideas about particles colliding, explain how the rate of the reaction will change as the temperature increases. [3]

(d) The Haber process is usually carried out at a higher temperature than the one which would give the highest yield. Suggest a reason for this. [1]

(MEG, Higher, 1998 Specimen)

13. Ammonia is made by combining nitrogen with hydrogen.

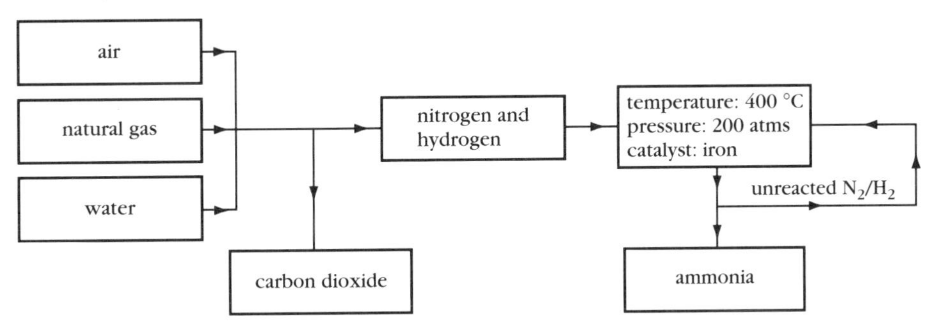

(a) (i) What are the **two** sources of hydrogen in this diagram? [1]

(ii) Carbon dioxide can be removed by reacting the gas with an alkaline solution. Explain why. [1]

(b) (i) Balance the equation for the formation of ammonia.

.... N_2 + ... $H_2 \rightleftharpoons$ .... NH_3 [1]

(ii) In this equation, what does $\rightleftharpoons$ represent? [1]

(iii) The rate of reaction to form ammonia is increased by increasing the pressure. Explain why, in terms of the collision theory. [2]

(iv) A catalyst is used in this reaction. The use of the catalyst makes the process more economical. Explain why. [3]

(c) Some of the ammonia is used to make the fertiliser ammonium nitrate.

$$\underset{\text{ammonia}}{NH_3} + \underset{\text{nitric acid}}{HNO_3} \longrightarrow \underset{\text{ammonium nitrate}}{NH_4NO_3}$$

(i) Calculate the relative formula mass of ammonium nitrate.
(Relative atomic masses are: H 1; N 14; O 16.) [1]

(ii) How many tonnes of ammonium nitrate can be made from 340 tonnes of ammonia? [3]

(iii) Explain **one** environmental problem caused by the excessive use of a fertiliser, such as ammonium nitrate. [2]

(SEG Double Award, Higher, 1998 Specimen)

Glossary

Acid Acids dissolve in water producing $H^+(aq)$ ions forming a solution with a pH of less than 7.

Acid rain Formed from acid gases (e.g. sulphur dioxide and nitrogen oxides) in the air dissolving in falling rain to produce an acidic solution.

Activation energy The excess energy that a reaction must acquire (E_A) to permit the reaction to occur.

Addition polymerisation The formation of a polymer by an addition reaction e.g. polyethene is formed from the monomer ethene.

Addition reaction A reaction in which an unsaturated compound becomes saturated.

Aerobic respiration The process by which living organisms take in oxygen from the atmosphere to oxidise their food to obtain energy.

Alcohols These are organic compounds containing the —OH group. They have the general formula $C_nH_{2n+1}OH$. Ethanol is by far the most important of the alcohols and is usually just called alcohol.

Alkali A soluble base which dissolves to produce $OH^-(aq)$ forming a solution of pH greater than 7.

Alkali metals The very reactive metals found in group 1 of the periodic table.

Alkaline earth metals The reactive metals found in group 2 of the periodic table.

Alkanes A family of saturated hydrocarbons with the general formula, C_nH_{2n+2}. The term saturated, in this context, is used to describe molecules that have only single covalent bonds.

Alkenes A family of unsaturated hydrocarbons with a general formula, C_nH_{2n}. The term 'unsaturated', in this context, is used to describe molecules which contain one or more double covalent carbon–carbon bonds.

Allotropes Different structural forms of the same element both having the same physical state. For example, carbon exists as the allotropes diamond, graphite and buckminsterfullerene.

Alloy A mixture of two or more metals. They are formed by mixing molten metals thoroughly.

Amphoteric hydroxide A hydroxide which can behave as an acid (react with an alkali) or a base (react with an acid), e.g. aluminium hydroxide.

Anaerobic respiration The process by which organisms obtain energy from chemically combined oxygen when they do not have access to free oxygen.

Anhydrous Being without water.

Anode The positive electrode.

Artificial fertiliser A substance which is added to soil to increase the amount of elements such as nitrogen, potassium and phosphorus. This enables crops grown in the soil to grow more healthily and to produce higher yields.

Atmosphere A mixture of gases that cloak a planet.

Atomic number (proton number) The number of protons in the nucleus of an atom. The number of electrons present in an atom. The position of the element within the periodic table. Given the symbol Z.

Atoms The smallest stable part of an element.

Avogadro's constant The number of particles in a mole of any substance, equal to 6.02×10^{23}.

Avogadro's law Equal volumes of all gases contain equal numbers of molecules under the same conditions of temperature and pressure.

Base A substance which neutralises an acid producing a salt and water as the only products.

Biodegradeable plastics These are plastics which have been designed to degrade (decompose) under the influence of bacteria, water or even sunlight.

Boiling point The temperature at which the vapour pressure of a liquid is equal to that of the atmosphere.

Bond A force which holds groups of atoms or ions together.

Bond energy The amount of energy required to break one mole of a given covalent bond.

Boyle's law At a constant temperature the volume of a given mass of gas is inversely proportional to the pressure.

$$V \propto \frac{1}{P}$$

Brownian motion Random motion of visible particles (pollen grains) caused by much smaller invisible ones (water).

Carboxylic acids A family of organic compounds containing the functional group —COOH. They have the general formula $C_nH_{2n+1}COOH$.

Catalyst A substance which alters the rate of a chemical reaction, usually speeding it up, without itself being chemically changed.

Catalytic cracking The decomposition of a higher alkane into alkenes and alkanes of lower relative molecular mass.

Cathode The negative electrode.

Centrifuging The separation of a mixture by rapid spinning. The denser particles are flung to the bottom of the containing tubes.

Ceramics A material made from inorganic chemicals by high temperature processing.

Charge Can be either positive or negative. Unlike charges attract whilst like charges repel.

Charles' law At a constant pressure the volume of a given mass of gas is directly proportional to the absolute temperature.

$$V \propto T$$

Chemical change It is a permanent change in which a new substance is formed.

Chemical equation A way of showing the changes which take place during a chemical reaction.

Chemical equilibrium It is dynamic. The concentrations of the reactants and products remain constant because the rate at which the forward reaction occurs is the same as that of the back reaction.

Chlor-alkali industry The industry based around the electrolysis of brine (saturated sodium chloride solution).

Chromatography A technique employed for the separation of mixtures of dissolved substances.

Combustion A chemical reaction in which a substance reacts rapidly with oxygen with the production of heat and light.

Competition reactions These are reactions in which metals compete for oxygen or anions. The more reactive metal: (i) takes the oxygen from the oxide of a less reactive metal, (ii) displaces the less reactive metal from a solution of that metal salt (a displacement reaction).

Complete combustion The complete combustion of a hydrocarbon molecule to form carbon dioxide and water only.
Composite material It is a material which combines the properties of the materials used to make it.
Compound A substance formed by the combination of two or more elements in fixed proportions.
Compress The particles within a substance may be squashed together by an increase in pressure.
Condense To change from a vapour or a gas into a liquid. The process is accompanied by the evolution of heat.
Contact process The industrial manufacture of sulphuric acid using the raw materials sulphur and air.
Continental crust The Earth's crust under the continents.
Continuous process A chemical process which is constantly fed with reactants.
Contraction Becoming smaller.
Convection currents The transference of heat through a liquid or gas by the actual movement of the substance.
Coulombs The unit used to measure the quantity of electricity.
Core The central part of the Earth, under the mantle, consisting mainly of nickel and iron.
Corrosion The name given to the process which takes place when metals and alloys are chemically attacked by oxygen, water or any other substances found in their immediate environment.
Covalent bond A chemical bond formed by the sharing of one or more pairs of electrons between two non-metal atoms.
Cross-linking This is the formation of side covalent bonds linking different polymer chains and therefore increasing the rigidity of the plastic. Thermosetting plastics are usually heavily cross-linked.
Crude oil A fossil fuel formed from the partly decayed remains of marine animals. Consists largely of hydrocarbons.
Crust The outer layers of the Earth.
Crystal lattice (ionic lattice) A regular 3D arrangement of atom/ions in a crystalline solid.
Crystallisation The process of forming crystals from a liquid.

Deliquescence The process by which water is absorbed by a substance from the surroundings to produce a concentrated solution.
Delocalised Becomes detached from the atom and spreads throughout the structure.
Density The mass per unit volume of a substance.
Deposition The laying down of sediments.
Desalination The production of fresh water from seawater.
Diatomic molecules A molecule containing two atoms e.g. hydrogen (H_2), oxygen (O_2).
Diffusion The process by which different substances mix as a result of the random motions of their particles.
Displacement reaction A reaction in which a more reactive element displaces a less reactive element from solution.
Dissolve To go into solution.
Distillation The process of boiling a liquid and then condensing the vapour produced back into a liquid. It is used to purify liquids and to separate mixtures of liquids.
Double covalent bond A chemical bond formed by the sharing of two pairs of electrons between two non-metal atoms.
Double decomposition The process by which an insoluble salt is prepared from solutions of two suitable soluble salts.
Ductile The property of a metal which enables it to be drawn out into a wire.
Dynamic equilibrium An equilibrium during a chemical reaction in which both the forward and back reactions occur at the same time.

Earthquakes A shaking of the Earth's crust caused by plates scraping past each other on a fault.
Effervescence Fizzing caused by bubbles of a gas being formed within a liquid during a chemical reaction.
Electrode A point where the electric current enters and leaves the electrolytic cell.
Electrolysis A process in which a chemical reaction is caused by the passage of an electric current.
Electrolyte A substance which will carry electric current only when it is molten or dissolved.
Electron A fundamental sub-atomic particle, with a negative charge, present in all atoms within energy levels around the nucleus.
Electron configuration (electron structure) A shorthand method of describing the arrangement of electrons within the energy levels of an atom.
Electron shells (energy levels) The allowed energies of electrons in atoms.
Electroplating The covering of objects with a thin layer of a metal using the process of electrolysis.
Electrostatic forces of attraction A strong force of attraction between unlike charges.
Element A substance which cannot be further divided into simpler substances by chemical methods.
Empirical formula This shows the simplest ratio of atoms present in a substance.
Endothermic reaction A chemical reaction which absorbs heat energy from its surroundings.
Enthalpy Energy stored in chemical bonds, given the symbol H.
Enthalpy change Given the symbol $\triangle H$ and represents the difference between energies of reactants and products.
Enthalpy change of neutralisation The enthalpy change that takes place when one mole of hydrogen ions is completely neutralised.
e.g. $H^+(aq) + OH^-(aq) \rightarrow H_2O(l)$
Enthalpy of combustion The enthalpy change which takes place when one mole of a substance is completely burned in oxygen.
Enzyme A protein molecule produced in living cells. It acts as a biological catalyst and is specific to a certain reaction. Enzymes only operate within narrow temperature and pH ranges.
Era A period of geological time considered as being of a distinctive character.
Erosion The wearing away of rocks and other deposits on the surface of the Earth by the action of wind, water and ice.
Ester A family of organic compounds formed by the reaction of an alcohol with a carboxylic acid in the presence of concentrated acid. This type of reaction is known as esterification.
Evaporation A process occurring at the surface of a liquid and involves the change of state of a liquid into a vapour at a temperature below the boiling point.

Exothermic reaction A chemical reaction in which heat energy is produced and released to its surroundings.
Expansion Becoming larger.
Extrusive igneous rocks Igneous rocks formed quickly when magma bursts through the Earth's crust.

Faraday One mole of electrons, which is equivalent to 96 500 coulombs.
Faults They are caused by the movement along fractures in the Earth's crust.
Fermentation A reaction in which sugar is changed into alcohol and carbon dioxide by the action of enzymes in yeast.
Fertiliser A substance which is added to the soil to increase its productivity.
Filtrate The liquid which passes through the filter paper during filtration.
Filtration The process of separating a solid from a liquid. It is done by the use of a fine filter paper which does not allow the solid to pass through.
Fissures Long, narrow cracks in rock.
Fossil fuels Fuels, such as coal, oil and natural gas, formed from the remains of plants and animals.
Fossils These are remains often found in sedimentary rocks of organisms deposited before the rock was formed.
Fractional distillation A method of separating a mixture of miscible liquids whose boiling points are quite close together.
Fractions These are the simpler mixtures or single substances obtained from fractional distillation.
Fuels Substances which produce heat energy when they are burned.
Functional group The atom or group of atoms responsible for the characteristic reactions of a compound.

Galvanising The process of coating a metal with zinc.
Gas The physical state in which the particles of a substance have so little attractive forces between them that no regular arrangement is seen.
Giant ionic structures A lattice held together by the electrostatic forces of attraction between ions.
Giant molecular substance A molecule containing thousands of atoms per molecule.
Glacier A slowly moving mass of ice.
Glass A hard, brittle substance made by heating together non-metal and metal oxides.
Greenhouse effect The absorption of infra-red radiation by gases such as carbon dioxide (a greenhouse gas) leading to atmospheric warming.
Groups A vertical column of the periodic table containing elements with similar properties with the same number of electrons in their outer energy levels. The elements have an increasing number of inner energy levels down the group.

Haber process The industrial manufacture of ammonia gas from nitrogen and hydrogen.
Halogens The non-metallic elements found in group 7 of the periodic table.
Hardness of water Caused by the presence in water of calcium (or magnesium) ions, which form a scum with soap and prevent the formation of a lather. There are two types of hardness.
Temporary hardness – caused by the presence of dissolved calcium (or magnesium) hydrogencarbonate.
Permanent hardness – this results mainly from dissolved calcium (or magnesium) sulphate.
Heat of reaction The energy given out or taken in when a chemical reaction occurs.
Homologous series A family of organic compounds which have similar structure, name endings and chemical properties. They show a trend in physical properties and their formulae can be represented by a general formula.
Hydrocarbon Is a molecule which contains atoms of carbon and hydrogen only.
Hydroelectric power Electricity generated by falling water turning turbines.

Igneous rocks Rocks formed when hot magma from the Earth's mantle cools and hardens. They are usually crystalline and are of two types, intrusive and extrusive. Intrusive igneous rocks e.g. granite, are formed by crystallisation of the magma underground. Extrusive igneous rocks, e.g. basalt, are formed by crystallisation of the magma on the Earth's surface.
Immiscible When liquids form two layers when mixed together they are said to be immiscible.
Incomplete combustion The incomplete oxidation of a hydrocarbon resulting in the formation of carbon monoxide and water.
Indicator A substance used to show whether a substance is acidic or alkaline (basic) e.g. phenolphthalein.
Inert gases Unreactive gaseous elements found in group 0 of the periodic table.
Inner core Solid rock at the centre of the Earth at very high temperature and pressure, composed of mainly nickel and iron. It has a diameter of 2530 km.
Insoluble If the solute does not dissolve in the solvent it is said to be insoluble.
Intermolecular bonds/forces (van der Waals' forces) Weak attractive forces which act between molecules, for example van der Waals' forces.
Intramolecular bonds Forces which act within a molecule e.g. covalent bonds.
Intrusive igneous rocks Igneous rocks formed by molten rock being forced up into the Earth's crust but not breaking the surface.
Ion An atom or group of atoms which has either lost one or more electrons making it positively charged or gained one or more electrons making it negatively charged.
Ion exchange The process in which one ion is exchanged for another ion.
Ionic bond A strong electrostatic force of attraction between oppositely charged ions.
Ionic equation The simplified equation which we can write if the chemicals involved are ionic.
Isomers Compounds which have the same molecular formula but different structural arrangements of the atoms.
Isotopes Atoms of the same element which possess different numbers of neutrons. They differ in mass number (nucleon number).

Kinetic theory A theory which accounts for the bulk properties of matter in terms of the constituent particles.

Lava Magma which has reached the surface of the Earth.
Lattice A regular 3-D arrangement of atoms/ions in a crystalline solid.

Law of conservation of mass For any chemical reaction the total mass of reactants is equal to the total mass of products produced.
Leaching The extraction of a soluble compound by dissolving it in a solvent.
Lime A white solid known chemically as calcium oxide (CaO). It is produced by heating limestone. It is used to cure soil acidity and to manufacture calcium hydroxide (slaked lime). It is also used as a drying agent in industry.
Limestone This is a form of calcium carbonate ($CaCO_3$). Other forms include chalk, calcite and marble. It is used directly to neutralise soil acidity, in the manufacture of iron and steel, glass, cement, concrete, sodium carbonate, and lime.
Liquid A state of matter intermediate between a solid and a gas.
Lubricant Any substance used to reduce friction between two moving surfaces.

Macromolecules Very large molecules.
Magma Is the molten rock material containing dissolved gases as well as water beneath the Earth's crust.
Magnetic Can be attracted to a magnet.
Malleable The property of a metal which allows it to be hammered into thin sheets without breaking.
Mantle A thick layer of solid, dense rock rich in magnesium and silicon which surrounds the outer core of the earth.
Mass The amount of matter in a substance, measured in kilograms.
Mass number (nucleon number) The total number of protons and neutrons found in the nucleus of an atom. Given the symbol A.
Matter Anything which occupies space and has a mass.
Melting point The temperature at which a solid begins to liquefy. Pure substances have a sharp melting point.
Membrane cell An electrolytic cell used for the production of sodium hydroxide, hydrogen and chlorine from brine in which the anode and cathode are separated by a membrane.
Metallic bond An electrostatic force of attraction between the mobile sea of electrons and the regular array of positive metal ions within the solid metal.
Metalloids (semi-metals) Any of the class of chemical elements intermediate in the properties between metals and non-metals, for example, boron and silicon.
Metals A class of chemical elements which have a characteristic lustrous appearance and which are good conductors of heat and electricity.
Metamorphic rocks These are rocks formed when rocks buried deep beneath the earth's surface are altered by the action of great heat and pressure. For example, marble is a metamorphic rock and is formed from limestone by this type of action.
Mineral A naturally occurring substance of which rocks are made.
Miscible When two liquids form one layer when mixed together they are said to be miscible.
Mixture A system of two or more substances which can be separated by physical means.
Molar mass The mass of one mole of a compound.
Molarity The number of moles of a solute dissolved in 1 dm^3 of solution.
Mole (mol) A mole of any substance contains 6×10^{23} atoms, ions or molecules. This number is called Avogadro's constant.
Molecular formula A formula showing the actual number of atoms of each element present.
Molecules Groups of atoms chemically bonded together.
Monatomic molecules A molecule which consists of only one atom e.g. neon, argon.
Monomer A simple molecule, such as ethene, which can be polymerised.

Negatively charged Having a negative electrical charge.
Neutral A substance with a pH of 7, neither acid nor alkaline.
Neutral (uncharged) Having no electrical charge.
Neutralise The process in which the acidity of a substance is destroyed. Destroying acidity means removing $H^+(aq)$ ions by reaction with a base, carbonate or metal.
Neutron A fundamental, uncharged sub-atomic particle present in the nuclei of atoms.
Nitrogen cycle The system by which nitrogen and its compounds both in the air and soil are interchanged.
Nitrogen fixation The direct use of atmospheric nitrogen in the formation of important compounds of nitrogen. Bacteria present in root nodules of certain plants are able to take nitrogen directly from the atmosphere to form essential protein molecules.
Nitrogenous fertiliser A substance which is added to soil to increase the amount of the element nitrogen.
Non-metals A class of chemical elements that are typically poor conductors of heat and electricity.
Non-renewable energy sources Sources of energy, such as fossil fuels, which take millions of years to form which we are using up at a rapid rate.
Nucleus The dense central part of an atom which contains the protons and neutrons.

Ocean-floor spreading The process which increases the width of an ocean by magma emerging and pushing older rock apart.
Oceanic crust The Earth's crust under the oceans.
Oil refining The general process of converting the mixture that is crude oil into separate fractions. These fractions, known as petroleum products, are used as fuels, lubricants, bitumens and waxes. The fractions are separated from the crude oil mixture by fractional distillation.
Optimum temperature A compromise temperature used in industry to ensure that the yield of product and the rate at which it is produced makes the process as economic as possible.
Ore A naturally occurring mineral from which a metal can be extracted.
Organic compounds Compounds which contain the element carbon often bonded to hydrogen and also to other elements such as oxygen, nitrogen, the halogens, sulphur and phosphorus.
Outer core Very dense liquid rock at high temperature composed of nickel and iron. The Earth's magnetic field arises here. It extends to a diameter of 6930 km.
Oxidation The addition of oxygen to, or the loss of electrons from, a substance.
Oxidising agent A substance which brings about oxidation.

Particles Extremely small pieces of matter, for example, atoms, molecules and ions.

Period A horizontal row of the periodic table. Within a period the atoms of all the elements have the same number of occupied energy levels but have an increasing number of electrons in the outer energy level.
Periodic table A table of elements arranged in order of increasing atomic number to show the similarities of the chemical elements with their related electron structures.
pH scale A scale running from 0–14 used for expressing the acidity or alkalinity of a solution.
Photosynthesis The chemical process by which green plants synthesise their carbon compounds from atmospheric carbon dioxide using light as the energy source and chlorophyll as the catalyst.
Planet An object which revolves around a star and which is illuminated by that star.
Plastics Most plastics are polymers and are classified as either thermosetting or thermosoftening materials.
Polymer A substance possessing very large molecules consisting of repeated units or monomers. Polymers, therefore have a very large relative molecular mass.
Polymerisation The chemical reaction in which molecules (the monomers) join together to form a polymer.
Positively charged Having a positive electrical charge.
Precipitate An insoluble compound formed from solution during a chemical reaction.
Precipitation The process of forming a precipitate.
Primary atmosphere The original thick layer of gases, mainly hydrogen and helium, that surrounded the core soon after the planet was formed 4500 million years ago.
Products Substances produced during a chemical reaction.
Proton A fundamental sub-atomic particle which has a positive charge, equal in magnitude to that of an electron. The proton occurs in all nuclei.

Quarrying Open surface excavation for the extraction of minerals or ores.

Radioactive These atoms break up spontaneously with the emission of certain types of radiation.
Rate of reaction The rate at which a reactant is used up or a product is formed in unit time.
Raw material The starting material from which the more refined substance is produced.
Reactants The starting materials in a chemical reaction.
Reactivity The property of a chemical which determines how readily it takes part in a chemical reaction.
Reactivity series An order of reactivity, giving the most reactive metal first, based on results from experiments with oxygen, water and dilute hydrochloric acid.
Redox A term applied to any chemical process which involves both reduction and oxidation.
Reducing agent A substance which brings about reduction.
Reduction The removal of oxygen from, or the addition of electrons to, a substance.
Relative atomic mass Symbol A_r.

$$A_r = \frac{\text{average mass of the isotopes of the element}}{1/12 \times \text{mass of a carbon 12 atom}}$$

Relative formula mass (relative molecular mass) This is the sum of the relative atomic masses of all those elements shown in the formula of the substance. This is often referred to as the relative molecular mass (M_r).
Renewable energy Sources of energy which cannot be used up, or which can be made at a rate faster than the rate of use.
Residue The solid left behind in the filter paper after filtration has taken place.
Respiration It is a chemical reaction in which glucose is broken down in a cell and energy is released. Respiration takes place in all living organisms at all times.
Reversible reaction A chemical reaction which is said to be reversible can go both ways. This means that once some of the products have been formed they will undergo a chemical change once more to reform the reactants. The reaction from left to right, as the equation for the reaction is written is known as the forward reaction and the reaction from right to left is known as the back reaction.
Rock cycle The cycle of natural rock change in which rocks are lifted, eroded, transported, deposited and possibly changed into another type of rock and then uplifted to start a new cycle.
Rusting Rust is a loose, orange/brown, flakey layer of hydrated iron(III) oxide found on the surface of iron or steel. The conditions necessary for rusting to take place are the presence of oxygen and water. The rusting process is encouraged by other substances such as salt. It is an oxidation process.

Salts Substances formed when the hydrogen of an acid is completely replaced by a metal or the ammonium ion (NH_4^+).
Salt hydrates Are the salts which contain water of crystallisation.
Saturated compounds Molecules which possess only single covalent bonds.
Saturated solution This is a solution which contains as much dissolved solute as it can at a particular temperature.
Secondary atmosphere Early volcanic activity created this mixture of gases. The mixture which formed this atmosphere included ammonia, nitrogen, methane, carbon monoxide, carbon dioxide and sulphur dioxide gases.
Sediment Matter that settles to the bottom of rivers, seas or oceans.
Sedimentary rock Rock which is formed when solid particles carried or transported in seas or rivers are deposited. Layers of sediment pile up over millions of years and the pressure created on the sediments at the bottom causes the grains to be cemented together.
Seismometer An instrument used to monitor the magnitude of earthquakes.
Silt Fine deposit of small particles of rock.
Simple molecular structures These substances possess between two and one hundred atoms per molecule.
Single covalent bond A chemical bond formed by the sharing of one pair of electrons between non-metal atoms.
Slaking The addition of water to lime.
Solid One of the three states of matter. In a solid the particles are arranged in a regular manner and they are only able to vibrate about a fixed position.
Soluble If the solute dissolves in the solvent it is said to be soluble.
Solubility The solubility of a solute in water at a given temperature is the number of grams of that solute which can dissolve in 100 g of water to produce a saturated solution at that temperature.
Solubility curve These are graphs of solubility against temperature.

Solute The substance which dissolves in the solvent.
Solution This is formed when a substance (solute) disappears into (dissolves) another substance (solvent).
Solvent The substance in which the solute dissolves.
Spectator ions Ions which remain unchanged during a chemical reaction.
States of matter Solid, liquid or gas.
Strong acid Is one which produces a high concentration of $H^+(aq)$ ions in water solution e.g. hydrochloric acid.
Strong alkali Is one which produces a high concentration of $OH^-(aq)$ ions in water solution e.g. sodium hydroxide.
Structural formula A formula which shows how the atoms in a molecule are arranged.
Sub-atomic particles Those particles found within an atom.
Sublimation The direct change of state from solid to gas and the reverse process.
Suspension A mixture in which visible solid particles do not settle under the force of gravity.

Tectonic plates The sections of the Earth's crust which move slowly about the surface of the Earth. The driving force behind the movement is thought to be convection currents in the mantle.
Tensile strength A measure of the ability of a material to withstand a stretching force.
Thermal dissociation (dissociation) The breakdown of a substance under the influence of heat.
Thermoplastics These are plastics which soften when heated (e.g. polyethene, PVC).
Thermosetting plastics These are plastics which do not soften on heating but only char and decompose (e.g. melamine).
Titration A method of volumetric analysis in which a volume of one reagent (usually an acid) is added to a known volume of another reagent (usually an alkali) slowly from a burette until an end-point is reached. If an acid and alkali are used then an indicator is used to show that the end-point has been reached.
Transition temperature The temperature boundary at which one allotropic form of an element is converted into another allotropic form.
Transition elements A block of metallic elements situated between groups 2 and 3 in the periodic table.

Ultraviolet radiation A high energy electromagnetic radiation just beyond the violet region of the visible spectrum.
Universal indicator A mixture of many other indicators. The colours shown by this indicator can be matched against the pH scale.
Unsaturated compound A compound which contains one or more double covalent bonds.

Valency The combining power of an atom or group of atoms. The valency of an ion is equal to its charge.
Vents The openings in the volcano through which volcanic gases, water vapour, carbon dioxide etc., rise to the surface.
Vulcanising This is the process which makes rubber harder and increases elasticity. This is done by the formation of sulphur bridges between the rubber chains.

Water cycle This cycle shows how water circulates around the Earth. The driving force behind the water cycle is the Sun.
Water of crystallisation Water incorporated into the structure of substances as they crystallise e.g. copper(II) sulphate pentahydrate ($CuSO_4.5H_2O$).
Weak acid Is one which produces a low concentration of $H^+(aq)$ in water solution e.g. ethanoic acid.
Weak alkali Is one which produces a low concentration of $OH^-(aq)$ ions in water solution e.g. ammonia solution.
Weathering The action of wind, rain and frost on rock which leads to its erosion.

Yeast Unicellular micro-organism capable of converting sugar into alcohol and carbon dioxide.
Yield The quantity of a product obtained in a chemical reaction.

Answers to examination questions

Section One

1 a) A = gas
B = liquid
C = solid

b) i) liquid gas

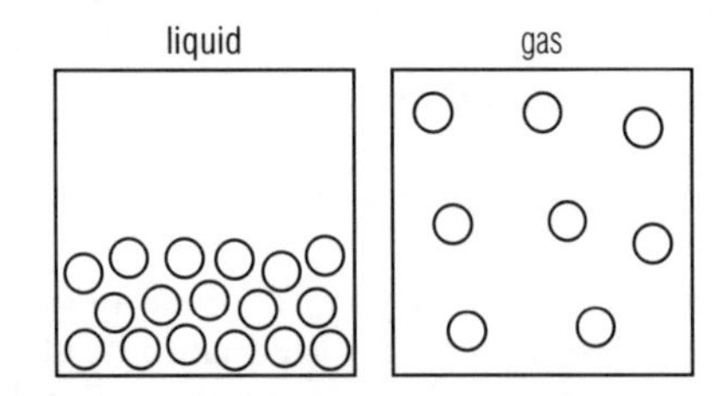

ii) When the solid metal is heated, the particles gain energy and vibrate more which causes them to move apart.

2 a) Elements – nitrogen, oxygen, argon
Compounds – carbon dioxide, water (vapour).

b) i) A mixture is more than one element and/or compound that is mixed physically but not chemically bonded together.
ii) A hydrocarbon is a substance which contains carbon and hydrogen only.

c) i) Oxygen
ii) Any two of carbon dioxide, carbon monoxide, water (vapour), oxides of nitrogen.

d) For example, carbon monoxide is poisonous and prevents blood transporting oxygen efficiently.
Or, carbon dioxide is a greenhouse gas and so causes a rise in the Earth's temperature. This will cause flooding.

3 a) A composite material is one which combines the properties of more than one material and so produces a more useful material.

b) Material – nylon
Reason – nylon is the weakest or least stiff (relative strength 0.8, relative stiffness 1.5).

c) Traditionally they are made from wood. Perhaps also the other materials would be socially/culturally unacceptable.

4 1st process – addition of water to dissolve the soluble salt (sodium chloride).
2nd process – stirring to speed up the formation of the solution.
3rd process – filtration to remove any insoluble solids.
4th process – evaporation to remove some of the water and so produce a very concentrated solution (saturated).
5th process – crystallisation used to separate the salt (sodium chloride) from the solution.

5 a) To ensure the salt is dissolved.

b) The filtrate is salt or sodium chloride.

c) It is washed to dissolve any remaining salt.

d) Evaporate the filtrate to dryness. The alternative would be to evaporate until crystals of salt appear and then leave to cool.

6 a) Magnesium – 24: 12 protons, 12 neutrons, 12 electrons.
Oxygen – 16: 8 protons, electron arrangement – 2.6.

b) The magnesium atoms lose 2 electrons to form the Mg^{2+} ion.
The oxygen atom gains 2 electrons to form the O^{2-} ion.

c) The attractive forces between the Mg^{2+} and the O^{2-} ions are much greater than between the Na^+ and Cl^- ions. So more energy is needed to overcome the attractive forces.

7 a) Either damp indicator paper is bleached or starch iodide paper turns blue-black.

b) Bromine, iodine – any element from group 7.

c) sodium + chlorine $\longrightarrow$ sodium chloride

d) The correct electron structure for sodium is 2.8.1 and for chlorine 2.8.7.

e) The sodium atom loses one electron to become a sodium ion (Na^+).

f) i) At the positive electrode – chlorine.
ii) At the negative electrode – hydrogen.
Chlorine is used in the manufacture of PVC, bleach, disinfectant.
Hydrogen is used in the manufacture of margarine and ammonia.
Sodium hydroxide is left in solution and used in the manufacture of ceramics and soap.

8 a) i) Some of the petrol would evaporate as it is being put into the tank. The particles have gained energy and spread out (diffused).
ii) Summer days are generally hotter. More heat energy is therefore available and so more particles will enter the vapour state (more will evaporate) and spread out.

b) The molecules in oil are much larger than the particles in petrol. There will therefore be stronger forces between these and so more energy will be needed to convert them from liquid to vapour.

c) Driving at quite high speed causes the tyres to warm up. At higher temperatures the particles of air in the tyre move faster. They make more collisions with the walls of the tyre. The pressure therefore increases at a constant volume.

9 a) liquid; **b)** gas; **c)** solid; **d)** gas.

10 i) 5; ii) B C, E; iii) CO_2.

11 a) 109

b) An element is a substance which cannot be split up any further into simpler substances by chemical means. These substances are made up of one type of atom only.

c) A compound is a substance which is made up of two or more elements chemically bonded together in a fixed proportion.

12 a) i) Metallic bonding diagram as shown in Section 1.11.
ii) Delocalised electrons can move through the metal, carrying a charge.
iii) The layers of metal ions will slide, when pulled, over each other.

b) i) Diagram of electron structure of Mg^{2+} and O^{2-} ions – as below.

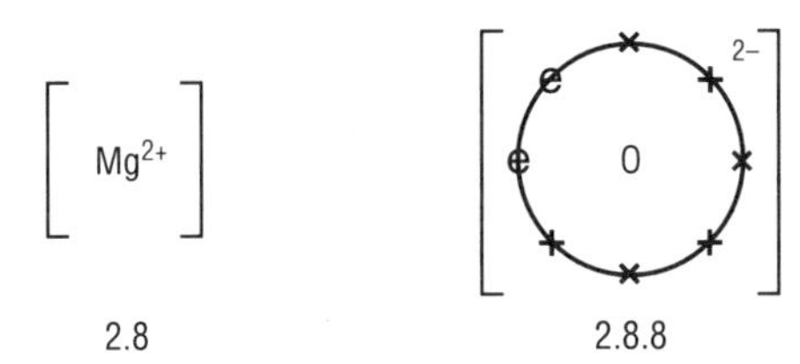

ii) It has a very high melting point and does not undergo thermal decomposition.

13a) i) 4 neutrons; 3 protons; 3 electrons.
ii) Isotopes
iii) Top – mass number = 6
Bottom – atomic number = 3
iv) Li^+

b) i) Electronic structure of 2, 8, 7.
ii) Diagram of Cl_2 molecule showing outer shell electrons only.

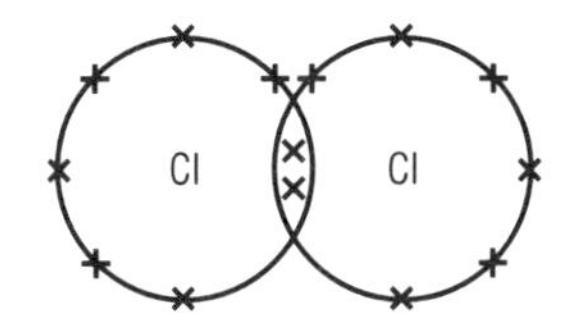

c) $Cl + e^- \longrightarrow Cl^-$
The chloride ion has one more electron so its electron structure is 2.8.8 (like argon).
The Cl atom has been reduced.

Section Two

1 a) No the statement is not the case since the modern periodic table is arranged in order of increasing atomic number. Yes it is the case that similar chemical properties do occur at intervals. These are the groups of the periodic table.

b) i) They have one electron in the outer shells (energy levels).
ii) The reactivity of these elements increases with increasing atomic number. The atoms get bigger and so it is easier to remove the outer electrons from the bigger atoms. There is more of a shielding effect from more inner shells.

c) i) Potassium floats on the water.
ii) The pH increases above 7 due to the production of potassium hydroxide, an alkali.
iii) Hydrogen gas is released and this is ignited by the heat produced from the reaction (exothermic).

d) i) All of these elements have full outer shells of electrons.
ii) Argon electron arrangement – 2.8.8.

2 a) i) Astatine is a solid.
ii) The most reactive halogen is chlorine.
iii) It is used because it kills harmful bacteria.

b) H 1, Cl 2.8.7, Na 2.8.1.

c) i) Diagram of the HCl molecule showing only the outer shells.

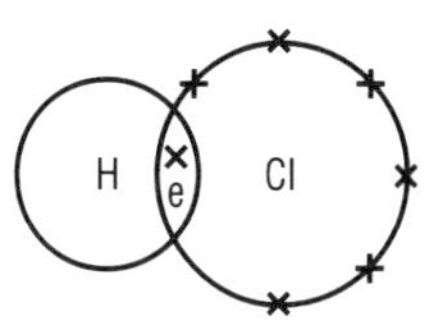

ii) $Na + Cl \longrightarrow Na^+ + Cl^-$
2.8.1 2.8.7 2.8 2.8.8

d) Giant structure because it has a high melting and boiling point, or it allows electricity to pass through the molten material.

3 a) The blue colour of the copper sulphate slowly changes to green and a pink-brown solid is seen.

b) $2Cr + 6HCl \longrightarrow 2CrCl_3 + 3H_2$

c) Carbon or carbon monoxide could be used as reducing agents. A high temperature is required.

d) i) The bumper will rust because Cr is less reactive than Fe. Therefore, the iron reacts in preference to the chromium.
ii) $Cr^{3+} + 3e^- \longrightarrow Cr$

4 a) i) An ore is a naturally occurring material from which a metal can be extracted.
ii) Haematite, coke, limestone, oxygen (from air).

b) i) Iron(III) oxide to iron is reduction.
ii) Oxygen is removed from the compound, *or* the iron ion gains electrons.
iii) Limestone is added to remove impurities.

c) i) Medium steel.
ii) It must be tough and flexible.

5 a) Because it contains a high level of calcium ion.

b) Calcium is good for bones and teeth.

c) You would expect to find caves containing stalactites and stalagmites.

d) Temporary hard water contains (calcium) hydrogencarbonate which can be removed by boiling. Permanent hard water contains, for example, calcium sulphate which cannot be removed by boiling.

6 a) i) The chemical breakdown (decomposition) of a substance by the passage of an electric current.
ii) Cathode.
iii) At the anode – bubbles of a green, pungent smelling gas are seen.
iv) At the cathode – bubbles of a colourless gas are seen.
v) $2H^+ + 2e^- \longrightarrow H_2$
(or $2H_2O + 2e^- \longrightarrow H_2 + 2OH^-$)

b) i) Impure copper is made the anode of the cell. Pure copper is made the cathode of the cell. The electrolyte is acidified copper sulphate solution. Pure copper will be deposited on the cathode.
ii) $Cu^{2+} + 2e^- \longrightarrow Cu$

7 a) The initial temperature of the acid and the volume of the acid used must be the same.

b) magnesium + hydrochloric acid $\longrightarrow$ magnesium chloride + hydrogen

c) Diagram of gas syringe apparatus.

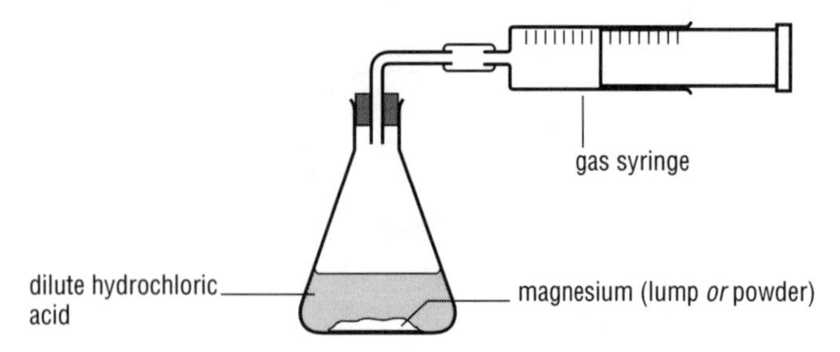

d) i) See graph.

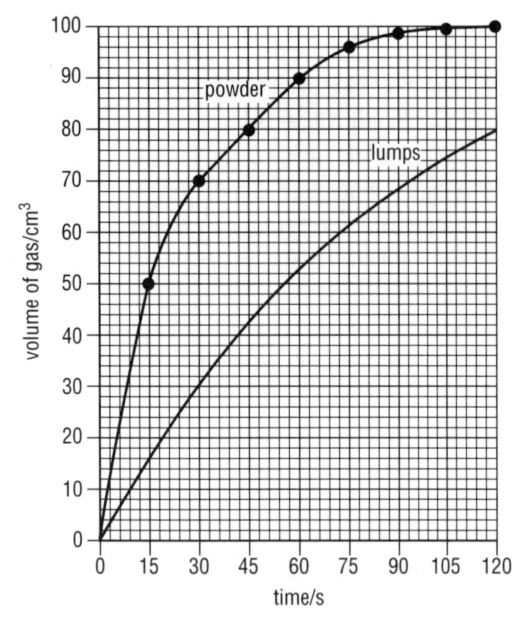

ii) The powdered magnesium reacts faster, giving hydrogen gas faster. Also it could be that the lumps have not finished reacting.
iii) A powder has a greater surface area and it is therefore possible for more acid particles to interact with the surface of the magnesium.
iv) In the powder reaction there is a faster reaction so more heat is given out and so raises the temperature of the acid. So the reaction becomes faster and faster.

8 a) Most reactive – C
E
A
D
Least reactive – B

b) Metal C would react too violently with steam or acid.

c) A = iron
B = gold
C = calcium
D = copper
E = magnesium

d) Metals have giant structures in which some of the outer shell electrons are free to move about through the whole structure – they are delocalised.

9 a) Aluminium is a reactive metal and so has reacted with oxygen to form aluminium oxide (the ore is bauxite – $Al_2O_3.2H_2O$).

b) i) The carbon anode.
ii) $Al^{3+} + 3e^- \longrightarrow Al$
iii) Oxygen
iv) During the process the level of the molten mixture goes down as the molten aluminium is siphoned off.
v) The molten aluminium has the higher density. It sinks to the bottom of the cell.

c) The equation is:

$$\text{coulombs} = \text{amps} \times \text{seconds}$$
$$= 200 \times 12 \times 60 \times 60 = 8\,640\,000\ \text{C}$$

The copper ion has a charge of $^{2+}$ and therefore needs 2 faradays to deposit 1 mole of copper.

$$Cu^{2+} + 2e^- \longrightarrow Cu$$

2 faradays is $2 \times 96\,500$ C and produces 64 g copper.
8 640 000 C will produce $\frac{8\,640\,000}{2 \times 96\,500} \times 64$ g copper
= 2865 g

10 a) Sodium hydroxide.

b) Neutralisation.

c) i) Universal indicator in pure water (pH = 7) is green.
ii) pH 3–4

d) The pH would increase.

e) i) Hydrogen ions (H^+).
ii) Hydroxide ions (OH^-).
iii) $H^+ + OH^- \longrightarrow H_2O$.

11 a) Add dilute silver nitrate solution to separate solutions of both sodium chloride and sugar. A white precipitate is produced with common salt (NaCl) but not with sugar in solution.

b) Dip a piece of universal indicator paper into both substances.
The acid turns the indicator paper pink whilst the water turns it green.

c) Add dilute hydrochloric acid to both samples. The carbonate will fizz (effervesce) but the sulphate is not affected.

12 a) i) This reaction is exothermic because more energy is liberated when O—H bonds in H_2O are formed than was required to break the H—H covalent bonds and O=O covalent bonds.
ii) This reaction does not happen because molecules do not have enough energy on their own to break the bonds of the substances.

b) The energy which is required to break the N—H covalent bonds is more than that liberated from the forming of the H—H and N≡N covalent bonds.

c) There are no bonds to break. So there is only energy liberated when P—P covalent bonds are made in P_4.

Section Three

1 a) i) Limestone or sandstone.
ii) Slate or marble.
iii) Granite or basalt.
b) Igneous.
c) Layers of sediment form and are subjected to pressure. This is due to the pressure from the layers themselves. At high pressure water and air are expelled (forced out) as the layers are compacted together into sedimentary rocks.
d) Granite is used for road making. Limestone is used in the manufacture of cement. Sand is used to produce concrete.

2 a) Marble, chalk.
b) i) Application of heat.
ii) Calcium hydroxide.
iii) [A] Turns chalky white (milky).
[B] It is used as a test for presence of CO_2.
c) Carbon dioxide is used in fizzy drinks as well as in fire extinguishers.

3 a) i) Crude oil was formed from organic matter (sea creatures) trapped under a build up of earthy materials over millions of years creating high temperatures and pressures.
ii) By drilling down to it.
iii) Crude oil is heated and the different compounds boil off at different temperatures.
iv) By heating large hydrocarbon molecules (alkanes) in the presence of a catalyst. These large molecules decompose to produce smaller molecules such as ethene.
b) i)
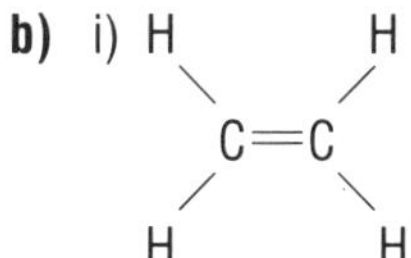

ii) Polymerisation.
iii) The double bonds open leaving a single bond. The molecules of ethene then join together by single bonds to form a very long polymer chain.
c) i) Methane – CH_4.
A_r values $12 + (4 \times 1) = 16$
ii) A_r value of C = 12
% of C in CH_4 is, $\frac{12}{16} \times 100 = 75\%$

4 a) i) It is a large molecule made from small molecules (monomers) joining (bonding) together.
ii) No air or oxygen is present.
iii) Used in alcoholic drinks or as a solvent or as a fuel.
iv) carbohydrate ⟶ FERMENTATION
$C_6H_{12}O_6$ 20–40 °C
v) Yeast.
b) i) Oxygen.
ii) It is acidic.
iii) Z is ethanoic acid.
c) i) The reaction removes the elements of water (H and O as H_2O).
ii) concentrated sulphuric acid
iii) ethanol ⇌ ethene + water
iv) The orange colour disappears and the solution becomes colourless.

5 a) C and E.
b) E.
c) Bicycle handlebars – A
Metal dustbins – B.
d) The solution could be evaporated.

6 a) i) Lowers cost and speeds up reaction rate.
ii) Sulphur dioxide + oxygen ⇌ sulphur trioxide.
b) i) Corrosive.
ii) Making detergents, fertilisers, fibres, paints, pigments, plastics.
c) i) It is a neutralisation reaction.
ii) The reaction works in the cold. The reaction rate is too violent/vigorous if heated.
iii) Copper(II) sulphate.

7 a) i) Protons and neutrons.
ii) The mass number (nucleon number)
$= 11p + 12n = 23$
b) The carbon-12 scale.
c)
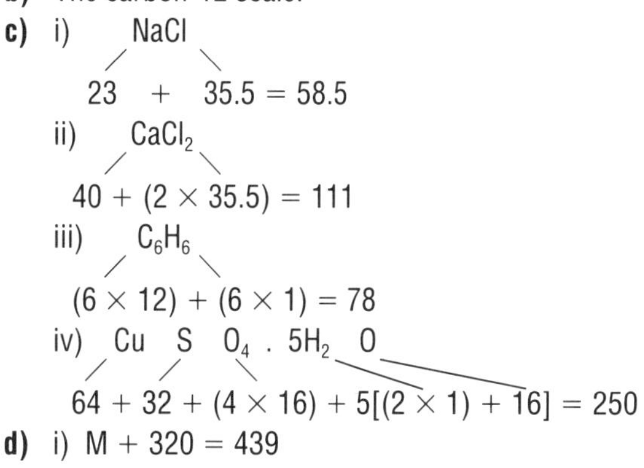

d) i) $M + 320 = 439$
$M = 439 - 320 = 119$.
ii) The element is tin.

8 a) Marks are gained for correct positioning of labels on diagram 2. A = N. America, B = Africa etc.
b) The land masses have shapes which fit closely. Also they have similar patterns of rocks and fossils.
c) The tectonic plates move due to the convection currents within the mantle caused by heat generated by radioactive processes.
d) The oceanic plate, on collision, is driven under the continental plate where it melts. The continental plates are then forced upwards.

9 a) When acid is dropped on limestone it fizzes.
b) i) 1 = calcium, 2 = carbon, 3 = oxygen
ii) Ca C O_3
$40 + 12 + 3(16) = 100$
c) At great pressure the water is squeezed out of the remains. These remains (minus water) are cemented together.
d) Limestone is buried deep underground and compressed at high pressure and temperature. This causes it to be converted into marble.

10 a) A compound made up of carbon and hydrogen only.
b) C_5H_{12}.

c) During fractional distillation crude oil is evaporated and the different fractions are condensed at different temperatures. Each fraction contains hydrocarbons with a similar number of carbon atoms. During cracking the larger, less useful hydrocarbon molecules are broken down into smaller alkanes which are more useful as fuels.

d) The molecule contains a double covalent bond.

e) 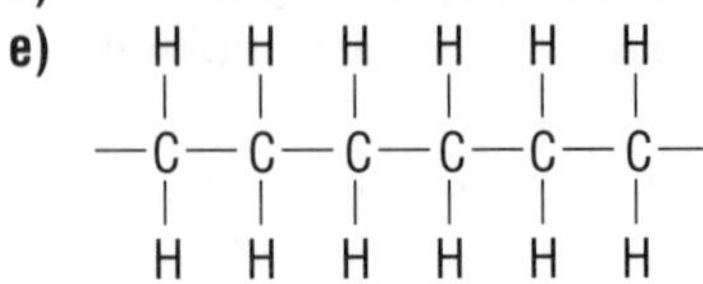

f) i) Polythene is unreactive; it; can be easily moulded; it is a solid in the right temperature range; it can be easily cleaned.
ii) When the thermosetting plastics are first heated they form 'cross-link' covalent bonds between adjacent chains. These 'cross-link' bonds are very strong and are broken on reheating.

11 a) i) l = liquid, g = gas.
ii) An unsaturated hydrocarbon contains a C=C (a double covalent bond between two C atoms).

b) i) Addition polymerisation involves a reaction in which a number of small unsaturated molecules (monomers) add to each other to form a very large molecule.
ii) $n(H_2C{=}CH_2) \longrightarrow -\!\!(\!-H_2C-CH_2-\!)_n\!-$
iii) Polythene is used for packaging as well as electrical insulation and for plastic containers.

c) i) B because it is flexible.
ii) C because it needs to be transparent.

12 a) It acts as a catalyst.

b) i) The yield of ammonia decreases.
ii) The yield of ammonia increases.

c) As the temperature increases the particles will move faster so more collisions will take place. Also they will be more energetic collisions. Therefore there is more likelihood of successful collisions in causing a reaction.

d) To maintain a faster rate of reaction.

13 a) i) Water and natural gas.
ii) Carbon dioxide is an acidic gas.

b) i) $N_2 + 3H_2 \rightleftharpoons 2NH_3$
ii) A reversible reaction.
iii) There are more particles per unit volume at higher pressures and hence there will be more collisions in a given time.
iv) The use of a catalyst reduces energy costs. Less pressure is therefore required by using a catalyst to speed up the reaction. Hence, lower machinery costs.

c) i) N H$_4$ N O$_3$

$14 + (4 \times 1) + 14 + (3 \times 16) = 80$

ii) From equation:
17 tonnes of ammonia ⟶ 80 tonnes of ammonium nitrate
so
340 tonnes of ammonia ⟶ $\frac{340 \times 80}{17}$ tonnes of ammonium nitrate
= 1600 tonnes

iii) In streams and rivers there will be excessive growth of photosynthetic organisms (eutrophication). This causes a high BOD when fish decay after suffocation. Other possibilities include the so-called 'blue baby syndrome' in babies. The ability of haemoglobin to carry oxygen is impaired by the presence of nitrite/nitrate.

Numerical answers to quick questions

Section 1.2

3 54.27 cm^3
4 23.81 cm^3
5 14.35 cm^3

Section 1.6

3 b) 20.2

Section 1.7

1 18 electrons

Section 2.8

1 a) 135 g/100 g water
b) 46 g/100 g water
2 a) i) 35 g/100 g water
ii) 44 g/100 g water
iii) 58 g/100 g water
b) i) 14 g (±0.50 g)
ii) 9 g (±0.50 g)
c) 3.30 g (±0.50 g)

Section 2.13

2 a) 2 **b)** 1 **c)** 3 **d)** 2
3 a) 193 000 coulombs
b) 96 500 coulombs
c) 193 000 coulombs
4 0.2 faradays
5 5594.2 s or 93 minutes 14.2 seconds
6 0.15 faradays

Section 2.14

2 c) 26 cm^3 (±0.5 cm^3)
d) 1 min 51 s (±3 s)

Section 2.15

1 a) −1299.50 kJ/mol
2 a) 364 kJ **b)** 182 kJ
3 a) 114 kJ **b)** 114 kJ

Section 3.10

2 b) 19.3 cm^3, 21.44%

Section 3.17

(p. 109)
1 a) 0.1 mol **b)** 0.17 mol
c) 2 mol
2 a) 3.20 g **b)** 160 g
c) 5.75 g
3 a) 0.1 mol **b)** 1 mol
c) 10 mol
4 a) 162 g **b)** 8.50 g
c) 56 g

(p. 111)
1 a) 0.08 mol
b) 0.0008 mol
2 a) 7.20 dm^3
b) 48 dm^3
3 a) 2 mol/dm^3
b) 0.2 mol/dm^3
4 7.98 g
5 0.224 M

(p. 113)
1 32 g
2 CaO

Chemistry data tables

Proton numbers, approximate relative atomic masses and properties of elements

Element	Symbol	Proton (Atomic) Number	Relative Atomic Mass (A_r)	M.Pt. in °C	B.Pt. in °C
Aluminium	Al	13	27	660	2470
Argon	Ar	18	40	−189	−186
Barium	Ba	56	137	725	1640
Bromine	Br	35	80	−7	59
Caesium	Cs	55	133	29	669
Calcium	Ca	20	40	840	1484
Carbon *Graphite *Diamond	C	6	12	sublimes sublimes	4800 4800
Chlorine	Cl	17	35.5	−101	−35
Copper	Cu	29	64	1084	2570
Fluorine	F	9	19	−220	−188
Germanium	Ge	32	73	937	2830
Gold	Au	79	197	1064	3080
Helium	He	2	4	−272	−269
Hydrogen	H	1	1	−259	−253
Iodine	I	53	127	144	184
Iron	Fe	26	56	1540	2750
Lead	Pb	82	207	327	1740
Lithium	Li	3	7	180	1340
Magnesium	Mg	12	24	650	1110
Mercury	Hg	80	201	−39	357
Nitrogen	N	7	14	−210	−196
Oxygen	O	8	16	−218	−183
Phosphorus	P	15	31	44	280
Potassium	K	19	39	63	760
Rubidium	Rb	37	85	39	686
Silicon	Si	14	28	1410	2355
Silver	Ag	47	108	960	2212
Sodium	Na	11	23	98	880
Strontium	Sr	38	88	769	1384
Sulphur (rhombic)	S	16	32	113	445
Tin (grey)	Sn	50	119	232	2270
Titanium	Ti	22	48	1660	3290
Zinc	Zn	30	65	420	907

* Carbon (as either graphite or diamond) does not melt but sublimes (i.e. changes directly from solid to gas); the temperature is approximate.

Indicators

Colour changes for universal indicator:

Colour:	Red	Orange	Yellow	Green	Blue	Navy blue	Purple
pH:	0–2	3–4	5–6	7	8–9	10–12	13–14

ACID NEUTRAL ALKALINE

Tests for ions

Ion	**Test**
Chloride (Cl^-)	Add a few drops of nitric acid and then a few drops of silver nitrate solution. A white precipitate is formed.
Sulphate (SO_4^{2-})	Add a few drops of hydrochloric acid and then a few drops of barium nitrate solution. A white precipitate is formed.
Carbonate (CO_3^{2-})	Add dilute hydrochloric acid to the solid (or mix with the solution). Bubbles of gas are given off.
Iron(II) (Fe^{2+})	Add aqueous sodium hydroxide to the solution. A pale green jelly-like precipitate is formed.
Iron(III) (Fe^{3+})	Add aqueous sodium hydroxide to the solution. A brown jelly-like precipitate is formed.
Copper(II) (Cu^{2+})	Add aqueous ammonia to the solution of copper(II) ions. A pale blue precipitate is formed; this dissolves when more ammonia solution is added and a deep blue solution is formed.

Periodic table of the elements

Key

Mass number A: 1
Proton number Z (Atomic number): 1
H hydrogen

Each cell: mass number, symbol, proton number, name.

Period	1 (I)	2 (II)											3 (III)	4 (IV)	5 (V)	6 (VI)	7 (VII)	0
1																		4 He 2 helium
2	7 Li 3 lithium	9 Be 4 beryllium											11 B 5 boron	12 C 6 carbon	14 N 7 nitrogen	16 O 8 oxygen	19 F 9 fluorine	20 Ne 10 neon
3	23 Na 11 sodium	24 Mg 12 magnesium											27 Al 13 aluminium	28 Si 14 silicon	31 P 15 phosphorus	32 S 16 sulphur	35.5 Cl 17 chlorine	40 Ar 18 argon
4	39 K 19 potassium	40 Ca 20 calcium	45 Sc 21 scandium	48 Ti 22 titanium	51 V 23 vanadium	52 Cr 24 chromium	55 Mn 25 manganese	56 Fe 26 iron	59 Co 27 cobalt	59 Ni 28 nickel	63.5 Cu 29 copper	65 Zn 30 zinc	70 Ga 31 gallium	73 Ge 32 germanium	75 As 33 arsenic	79 Se 34 selenium	80 Br 35 bromine	84 Kr 36 krypton
5	85 Rb 37 rubidium	88 Sr 38 strontium	89 Y 39 yttrium	91 Zr 40 zirconium	93 Nb 41 niobium	96 Mo 42 molybdenum	99 Tc 43 technetium	101 Ru 44 ruthenium	103 Rh 45 rhodium	106 Pd 46 palladium	108 Ag 47 silver	112 Cd 48 cadmium	115 In 49 indium	119 Sn 50 tin	122 Sb 51 antimony	128 Te 52 tellurium	127 I 53 iodine	131 Xe 54 xenon
6	133 Cs 55 caesium	137 Ba 56 barium	●	178.5 Hf 72 hafnium	181 Ta 73 tantalum	184 W 74 tungsten	186 Re 75 rhenium	190 Os 76 osmium	192 Ir 77 iridium	195 Pt 78 platinum	197 Au 79 gold	201 Hg 80 mercury	204 Tl 81 thallium	207 Pb 82 lead	209 Bi 83 bismuth	209 Po 84 polonium	210 At 85 astatine	222 Rn 86 radon
7	233 Fr 87 francium	226 Ra 88 radium	●	261 Rf 104 rutherfordium	262 Db 105 dubnium	263 Sg 106 seaborgium	262 Bh 107 bohrium	Hs 108 hassium	Mt 109 meitnerium	* 110	* 111	* 112						

* no names yet

139 La 57 lanthanum	140 Ce 58 cerium	141 Pr 59 praseodymium	144 Nd 60 neodymium	147 Pm 61 promethium	150 Sm 62 samarium	152 Eu 63 europium	157 Gd 64 gadolinium	159 Tb 65 terbium	162 Dy 66 dysprosium	165 Ho 67 holmium	167 Er 68 erbium	169 Tm 69 thulium	173 Yb 70 ytterbium	175 Lu 71 lutetium
227 Ac 89 actinium	232 Th 90 thorium	231 Pa 91 protactinium	238 U 92 uranium	237 Np 93 neptunium	244 Pu 94 plutonium	243 Am 95 americium	247 Cm 96 curium	247 Bk 97 berkelium	251 Cf 98 californium	252 Es 99 einsteinium	257 Fm 100 fermium	258 Md 101 mendelevium	259 No 102 nobelium	260 Lw 103 lawrencium

Properties of some common substances (under normal conditions)

Name	Melting Pt. in °C	Boiling Pt. in °C	Electrical conductivity (molten)	Structure
Aluminium oxide	2072	2980	good	giant
Ammonia	−77	−34	poor	molecular
Barium chloride	963	1560	good	giant
Calcium chloride	782	1600	good	giant
Calcium oxide	2614	2850	good	giant
Carbon dioxide	sublimes	−78	poor	molecular
Carbon monoxide	−199	−191	poor	molecular
Copper(II) chloride	620	993	good	giant
Copper(II) sulphate	200	decomposes	–	giant
Hydrogen chloride	−114	−85	poor	molecular
Iron(III) oxide	1565	–	good	giant
Lead chloride	500	950	good	giant
Lithium chloride	605	1340	good	giant
Lubricating oil	–	250–350	poor	molecular
Magnesium chloride	714	1412	good	giant
Methane (natural gas)	−182	−161	poor	molecular
Methylated spirit	−100	80	poor	molecular
Paraffin wax	50–60	decomposes	poor	molecular
Petrol	−60 to −40	40 to 75	poor	molecular
Potassium chloride	770	1500*	good	giant
Silicon dioxide (sand)	1610	2230	poor	giant
Sodium chloride	801	1413	good	giant
Sugar (sucrose)	161	decomposes	poor	molecular
Sulphur dioxide	−73	−10	poor	molecular
Water	0	100	poor	molecular

* Approximate value

Formulae of some common ions

Positive ions Name	**Formula**	**Negative ions Name**	**Formula**
Hydrogen	H^{+}	Chloride	Cl^{-}
Sodium	Na^{+}	Bromide	Br^{-}
Silver	Ag^{+}	Fluoride	F^{-}
Potassium	K^{+}	Iodide	I^{-}
Lithium	Li^{+}	Hydrogencarbonate	HCO_3^{-}
Ammonium	NH_4^{+}	Hydroxide	OH^{-}
Barium	Ba^{2+}	Nitrate	NO_3^{-}
Calcium	Ca^{2+}	Oxide	O^{2-}
Copper(II)	Cu^{2+}	Sulphide	S^{2-}
Magnesium	Mg^{2+}	Sulphate	SO_4^{2-}
Zinc	Zn^{2+}	Carbonate	CO_3^{2-}
Lead	Pb^{2+}		
Iron(II)	Fe^{2+}		
Iron(III)	Fe^{3+}		
Aluminium	Al^{3+}		

Reactivity series of metals

Potassium	most reactive
Sodium	
Calcium	
Magnesium	
Aluminium	
Carbon	
Zinc	
Iron	
Tin	
Lead	
Hydrogen	
Copper	
Silver	
Gold	
Platinum	least reactive

(Elements in *italics*, though non-metals, have been included for comparison.)

Common ores of metals

Name of ore	Chemical formula	Metal extracted
Haematite	Fe_2O_3	Iron
Bauxite	Al_2O_3	Aluminium
Halite	NaCl	Sodium
Zinc blende (sphalerite)	ZnS	Zinc
Malachite	$Cu_2CO_3(OH)_2$	Copper
Copper pyrites (chalcopyrite)	$CuFeS_2$	Copper
Ilmenite	$FeTiO_3$	Titanium
Wolframite	$FeWO_4$	Tungsten
Cassiterite	SnO_2	Tin
Galena	PbS	Lead

Pattern of magnetic 'stripes' on the ocean floor

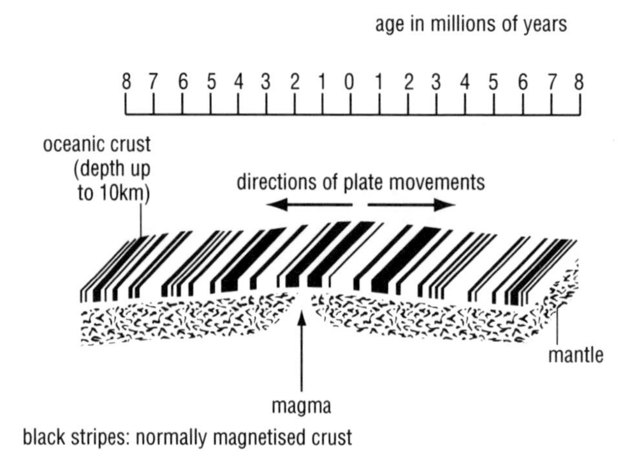

black stripes: normally magnetised crust

Identification of rock types

Typical sizes of sedimentary grains

Pebbles	4 to 64 mm
Grit	2 to 4 mm
Sand	0.06 to 2 mm
Mud/silt	0.004 to 0.06 mm

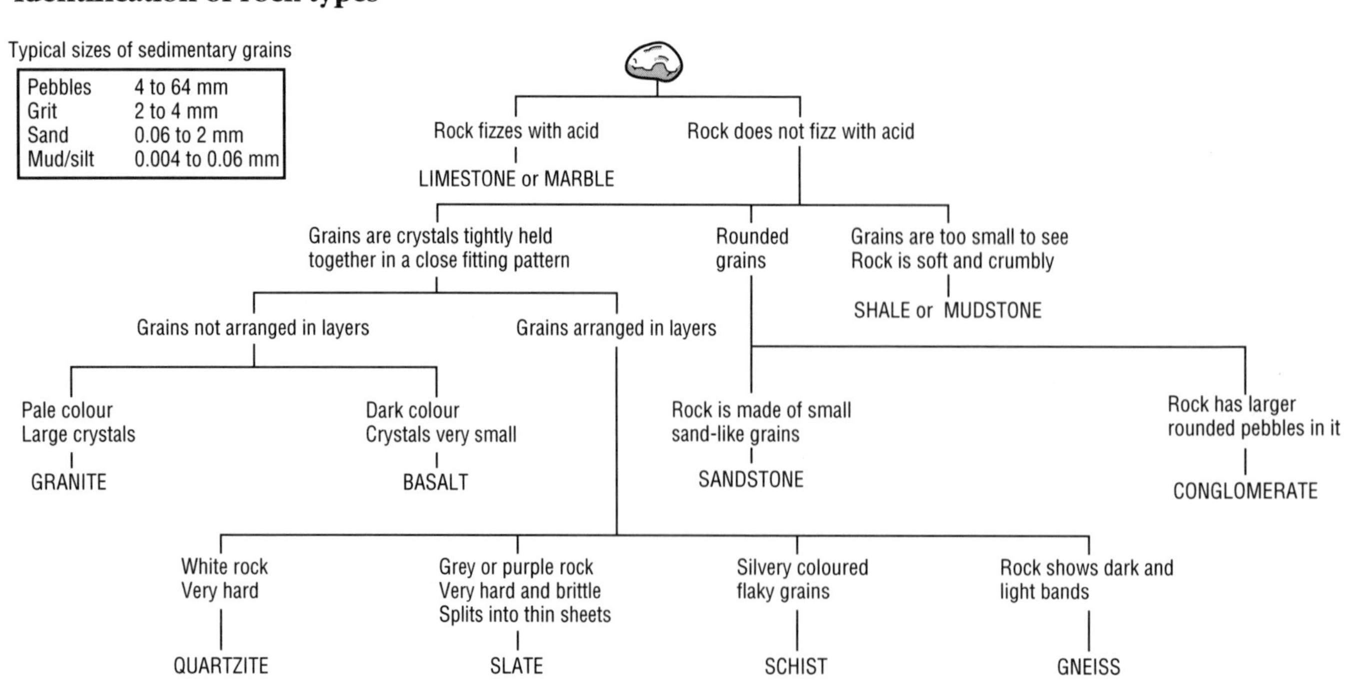

Index